NANOTECHNOLOGY
Ethics and Society

PERSPECTIVES IN NANOTECHNOLOGY

Series Editor
Gabor L. Hornyak

NANOTECHNOLOGY
Ethics and Society

Deb Bennett-Woods

CRC Press
Taylor & Francis Group
Boca Raton   London   New York

CRC Press is an imprint of the
Taylor & Francis Group, an **informa** business

CRC Press
Taylor & Francis Group
6000 Broken Sound Parkway NW, Suite 300
Boca Raton, FL 33487-2742

© 2008 by Taylor & Francis Group, LLC
CRC Press is an imprint of Taylor & Francis Group, an Informa business

No claim to original U.S. Government works

ISBN 13: 978-1-4200-5352-4 (pbk)

Library of Congress Cataloging-in-Publication Data

Nanotechnology : ethics and society / editor, Deb Bennet-Woods.
    p. cm. -- (Perspectives in nanotechnology)
    "A CRC title."
    Includes bibliographical references and index.
    ISBN 978-1-4200-5352-4 (alk. paper)
    1. Nanotechnology--Social aspects. 2. Nanotechnology--Moral and ethical aspects. I. Bennet-Woods, Deb.

T174.7.N373194 2008
303.48'3--dc22   2008009174

Visit the Taylor & Francis Web site at
http://www.taylorandfrancis.com

and the CRC Press Web site at
http://www.crcpress.com

Contents

PART ONE   Foundations

PART TWO   Emerging Issues

PART THREE   The Framework Applied

Series Foreword

Welcome to the Perspectives in Nanotechnology Series—a group of short, readable paperback books dedicated to expanding your knowledge about a new and exciting technology. The book you are about to read involves subject matter that goes beyond the laboratory and the production line. It is not about technical details—the book you have taken aboard your connecting flight, commuter train or bus, or to your hotel room involves a specific aspect of nanotechnology that will have some impact on your life, the welfare of your family, and the wealth and security of this nation. The degree of this impact may be unnoticeable, slight, overwhelming, or any place in between those extremes depending on the specific application, its magnitude and the scope of its distribution. Those of us who are able to recognize trends, conduct efficient research, plan ahead and adapt will succeed in a new world enhanced by nanotechnology. This book in the Perspectives in Nanotechnology Series hopefully will act as the catalyst for your *fantastic journey.*

Each book in the Series focuses on a selected aspect of nanotechnology. No technology exists in a vacuum. All technology is framed within the contexts of societal interactions, laws, and practices. Once a technology is introduced to a society, the society must deal with it. The impact of a technology on culture, politics, education, and economics depends on many complex factors—just reflect for a moment on the consequences (good and bad) of the computer, the automobile, or the atomic bomb. Nanotechnology is designated to be the "next industrial revolution." Although there is much hype associated with nanotechnology, the ability to manipulate atoms and molecules in order to fabricate new materials and devices that possess remarkable properties and functions alone should be enough of a hook to draw you in.

The impact of new technology is more relevant than ever. Consider that our world is highly integrated, communication occurs instantaneously and that powerful geopolitical and economic pressures are in the process of continually changing the global landscape. We repeat—the degree of the impact of nanotechnology may be unnoticeable, slight, overwhelming, or anyplace in between. Those of us who are able to recognize trends, conduct efficient research, plan ahead, and adapt will succeed. It is all about survival. It always has been. Darlene Geis in her book, *Dinosaurs and Other Prehisoric Animals,* states:

> ...and finally even the mighty T-Rex died out, too. His size and strength
> and remarkable jaws were of no use to him in a world that was changing
> and where his food supply was slowly disappearing. In the end, the king
> was no greater than his subjects in a world whose rule has always been
> Change with Me—or Perish![1]

Although stated with a bit of drama, the quotation does bring the point across quite effectively. Your future is in your hands—perhaps holding this very book.

Societal Implications. Societal aspects (implications) consist of a broad family of highly integrated components and forces that merge with technology to form our civilization. Government, business, academia, and other social institutions have evolved over millennia and are in a constant state of dynamic flux. Civilizations change for many reasons. Technology always has been one of the primary drivers of this change. The change may be beneficial, detrimental, or anywhere in between. From the first stone implement, the iron of the Hittites to the microchip, technology has always played a major role in the shaping of society. Societal implications of nanotechnology are rooted in the technology. Societal implications in turn have the capacity to alter any technology. How many times have social forces inspired a new technology? The technology developed in the space program is one example of such a relationship—the development of penicillin, another.

What exactly are "societal implications"? How do they relate to nanotechnology? In this series, we intend to cover a wide variety of topics. Societal implications of nanotechnology are numerous and diverse and encompass the legal, ethical, cultural, medical, and environmental disciplines. National security, education, workforce development, economic policy, public policy, public perception, and regulation are but a few of the areas we plan to address in the near future.[1,2] All aspects of government, business, and academia are subject to the influence of nanotechnology. All vertical industrial sectors will be impacted by nanotechnology—aerospace, health care, transportation, electronics and computing, telecommunications, biotechnology, agriculture, construction, and energy. For example, Fortune 500 companies already have staked a claim in nanotechnology-based products. Service industries that focus on intellectual property and technology transfer, health and safety, environmental management and consulting, workforce sourcing and job placement, education development and curriculum, and investment and trading already engage the challenges brought about by nanotechnology. There is no lack of subject matter. We plan to cover the most urgent, the most relevant, and the most interesting topics.

Ethical implications are associated with every form of technology. Artificial intelligence, weapons systems, life-extending drugs, surveillance, altered organisms, and social justice all have built-in moral implications— ready for us to discuss. Nanotechnology is creating new ethical dilemmas while simultaneously exacerbating (or alleviating) older ones. Nanotechnology is already changing our legal system. How does one go about obtaining a patent of a process or material that is the result of an interdisciplinary collaboration, e.g., the convergence of engineering, chemistry, physics, and biology? Even more so, the environmental footprint of nanotechnology is expected to be three orders of magnitude less than that of any current technology. The health (and environmental) consequences of nanomaterials are mostly unknown. And what of public perception? How many of you want a nanotech research center in your backyard (are you a NIMBY)? How should we update our educational system to accommodate nanoechnological topics? What should we do to make sure our workforce eis current and prepared? How will your job or career be influenced by nanotechnology?

There are other relevant questions. How does one go about building a nanobusiness? What new kinds of partnerships are required to start a business, and what exactly is the *barrier of entry* for such an undertaking? What are *nanoeconomic clusters*? What Fortune 500 companies and what business sectors require a book in this Series to describe its NT profile? And what of investing and funding? What is the status of nanotechnology programs on the international stage? What about nanotechnology and religion? What about the *future of nanotechnology*? The list goes on.

The Books. Web resources that address societal implications of nanotechnology are plentiful but usually offer encapsulated or cursory information. On the other hand, comprehensive (but tedious) summary reports produced by research and marketing firms are suitable for the serious investor, but require a major financial commitment to procure and, therefore, are generally not available to the public at large. In addition, government entities, e.g., the National Nanotechnology Initiative (www.nano.gov), have generated comprehensive reports on the societal impact of NT.[1, 2] Such documents, although excellent, are generally not well known to the public sector. A reader-friendly, affordable book with commercial appeal that targets the nano-aware (as well as the unaware) layperson or expert in the field offers a convenient alternative to the options listed above.

The intent of each book is to be informative, compelling, and relevant. The books, in general, adhere to the criteria listed below.

- **Readability.** Each book is 200 to 300 pages long, with easy-to-read font and abundant with non-technical, but certainly non-ponderous language.

- **References.** Each book is well researched and provides links to more detailed sources when required.

- **Economical Pricing.** Each book is priced within easy reach and designed for accelerated distribution at conferences and other venues.

- **The Subject Matter.** The subject of each book is relevant to nanotechnology and represents the cutting-edge in the state-of-the-art.

- **Relevance.** The books are dynamic. We must stay current if we are to abide by T-Rex's rule! Specifically, the content will stay relevant in the form of future editions as the climate of nanotechnology is expected to change dynamically over the years to come. A strong temporal component is inherent in the Perspectives in Nanotechnology Series.

It is our hope that readers delve into a book about their special interest, but also to transform themselves into a state of *nano-readines*. Are you nano-ready?

Do you want to be able to recognize the drivers that surround nanotechnology and its potential promise? Do you want to be able to learn about the science, technology, and potential implications? Are you ready at this time to plan and adapt to the changes? Do you want to become an agent of change? Do you want success in that future? If your answers are, in order—NO, YES, YES, NO, YES, and YES—you are ready to begin reading this book.

Gabor L. Hornyak, Ph.D.
Series Editor

___

References

1. Geis, D. 1959. *Dinosaurs and Other Prehistoric Animals*. Grosset & Dunlap, New York.
2. Roco, M. C., W. S. Bainbridge, eds. 2001. *Societal Implications of Nanoscience and Nanotechnology*. National Science Foundation, Arlington, Virginia.
3. Roco, M.C., W.S. Bainbridge, eds. *Nanotechnology: Societal Implications—Maximizing Benefits for Humanity*, Report of the National Nanotechnology Initiative Workshop, December 2-3, 2003.

Foreword

Michele Mekel

The nanotechnology boom—or bubble, depending upon whom one believes—is presently taking the world by storm. Billions of public dollars have been, and continue to be, spent by nations across the globe on the research and development of nanoscale technologies that promise to revolutionize almost every industrial sector. Moreover, "nano" is well on its way to becoming a household phrase due to its being appended to the names of consumer products ranging from pants to car wax, and from MP3 players to bathroom cleaners—regardless of whether or not these products actually contain nanoparticles. Yet, as the "nano" marketing machine heads into overdrive, study after study and poll after poll continue to show that the general public remains clueless as to even a basic definition of nanotechnology (Macoubrie, 2005; Scheufele, 2007).[1,2]

In the same vein as the "nano" marketing fad, another nano-craze phenomenon has emerged as well: "nano" parlance, which is the addition of the "nano" prefix to an existing word. And, of all the novel "nano-ese" expressions currently in vogue, one has drawn particular attention. That term is "nanoethics," which, at least in ivory-tower science and technology studies (STS) circles, already has been embraced as a distinct, stand-alone subset of ethical analysis in and of itself by some (Keiper, 2007),[3] qualified as perhaps too limited a concept by a few (Mekel and Cameron, 2007),[4] and derided as a strawman by others (Keiper, 2007).[3]

At its core, however, nanoethics (as well as its variants, including NELSI [nanotechnology's ethical, legal, and societal implications], NE³LS, SEIN, and NELS) (Mekel, 2006)[6] is simply a shorthand turn of phrase that encompasses the development, study, practice, and enforcement of a set of culturally accepted beliefs, mores, guidelines, standards, regulations, and even laws for governing rapidly advancing nanotechnologies across multiple economic sectors. While lagging behind nanoscale science and technology, attention to the identification, crystallization, formation, and implementation of these social values has recently been gaining momentum, albeit mainly in limited pockets, including the STS crowd, environmentally focused NGOs, emerging technology law and policy wonks, and intellectual property lawyers—with the broader public still very much out of the loop.

Given the expansive reach of nanotechnologies, however, the application of nanoscale science, which deals with the manipulation of the minutest components of matter (National Nanotechnology Initiative, n/d),[7] will likely, in time, give rise to culturally challenging issues of tremendous proportion and grave importance. As such, there is nothing "nano" about nanoethics. In fact, if nanotechnology is to live up to the big promises made by its proponents, then nanoethics must develop concurrently and in lock step with technological innovation in this arena. Only this coupling of scientific advancement and social values will ensure that nanotechnology, which is comprised of an extensive range of today's "leading-edge" technologies, does not become synonymous with the "bleeding edge" of technology.

Ambitiously tackling this topic and its nuances across sectors, Debra Bennett-Woods, author of this volume in the *Perspectives on Nanotechnology* series, brings the perfect balance and the right voice to bear on this complex, multi-faceted subject matter. An established ethicist and health care leader, Bennett-Woods is an expert at bridging the gaps between the pristine realm of academic theory and the not-so-pristine real world, where ideas must be tested and applied, often times, in very difficult and unforeseen situations. In so doing, she not only helps frame and advance this much-needed dialogue, but she does so in a way that makes the nanoethics discourse accessible to a broad audience.

Michele Mekel, J.D., M.B.A., M.H.A.
Michele Mekel is the Associate Director and Senior Fellow at the Center on Nanotechnology and Society of the Chicago-Kent College of Law/Illinois Institute of Technology.

References

1. Macoubrie, J. 2005. *Informed public perceptions of nanotechnology and trust in government.* Woodrow Wilson International Center for Scholars. Available at http://wilsoncenter.org/news/docs/Macoubriereport1.pdf
2. Scheufele, D. The public as consumers: Understanding attitudes toward nanotech (presentation at the Center for Nanotechnology and Society's Chicago Nano Forum, March 21, 2007, at Illinois Institute of Technology), available at: http://www.nano-and-society.org/events/past_events.html
3. Keiper, A. "Nanoethics as a discipline?" *The New Atlantis*, spring 2007: pp. 55-67. See also, The Nanoethics Group website, http://www.nanoethics.org/whynanoethics.html
4. Mekel, M. and N. Cameron. 2007. "Taking Stock of the NELSI Landscape." In N. Cameron and M. E. Mitchell, eds. *Nanoscale: Issues and Perspectives for the Nanocentury.* London: John Wiley.
5. Keiper, A. "Nanoethics as a discipline?" *The New Atlantis*, spring 2007: pp. 55-67.
6. Mekel, M. "Nanotechnologies: Small science, big potential and bigger issues." *Development*, 2006: 29(4):47-53.
7. National Nanotechnology Initiative website, What is nanotechnology?, available at: http://www.nano.gov/html/facts/whatIsNano.html

Preface

"Nanotechnology" is a term used to describe a broad category of technologies that involve manipulation of matter and its novel properties at the atomic and molecular levels. As an enabling technology, nanotechnology has the potential to alter every aspect of the human condition within the coming decades. Predictions of what, when, and how this will occur vary widely, but all agree there is tremendous possibility for both positive impacts and untoward consequences.

From manufacturing, to national security, to advances in medicine, nanotechnology implies revolutionary change. However, the sweeping changes wrought by a technological advance of this potential magnitude are likely to come at a price that includes unforeseen environmental impact, disruptions in industry, displacement of workers, and deeply controversial applications of the technology and its offspring. Technology has often been a double-edged sword, with new technologies solving one problem while creating one or more new problems. Furthermore, new technologies tend to be in use, often widespread, long before we have had the opportunity to fully consider and prepare for the consequences.

So, how do society and its various stakeholders assess and plan for the impacts of the next big technological revolution? How do we reconcile competing economic, political, cultural, social, and environmental priorities when thinking through the complex and somewhat unpredictable interactions that accompany rapid technological advancement? How do we weigh the relative roles and responsibilities of scientists, engineers, entrepreneurs, corporations, policy makers, regulators, community leaders, and individual citizens when evaluating short- and long-term implications? And finally, how do we accomplish any of this in the face of the increasingly rapid pace of evolving science and emerging technology?

While these questions are broadly applicable to any technology, they are particularly appropriate in the current societal context. Consider an image of society as a complex system of boundaries rather than a monolithic whole. The term "boundaries" represents the myriad of competing interests, priorities, values, and perceptions that make up our social reality and its dynamic interface with science and technology. Social impact happens when a boundary is crossed in a way that alters some aspect of the larger human condition. Ethical issues serve as a particularly powerful lens for examining and evaluating the nature of the impact, and then directing our individual and collective responses.

Society's boundaries are best represented by the voices of those persons embedded within them. Therefore, in an effort to provide an appropriately balanced and representative sample of the voices engaged in the emerging nanotechnology enterprise, each chapter of this book contains one or more brief commentaries by a range of nanotechnology stakeholders. These voices speak to the compelling nature of the benefits and burdens that accompany emerging technologies as well as the tremendous opportunity for reasoned dialogue.

In this spirit, the purpose of this book is to provide a language, a conceptually clear and straightforward framework, in which pragmatic questions can be raised regarding the impact of nano-related technologies. It is my hope this common language can be applied by anyone—from students to scientists and engineers, business leaders, policy makers, and the general public—to establish a common ground for reflection, dialogue, and action.

While a book of this nature cannot possibly cover all of the potential issues arising from advancements in nanotechnology, it can employ examples of applications that are likely to have broad reaching effects in our every day lives. In order to raise awareness and illustrate the process of ethical analysis and dialogue, the book will focus on general issues related to nanotechnology in nanomaterials and manufacturing as well as impacts on the marketplace and workforce. In addition, the military and national security, the environment, and health and medicine will be considered in more detail as special cases to illustrate the breadth of issues raised by nanotechnology and its enabling effects.

This book is not a philosophical treatise on science and technology. Rather, the book is written with the intent of provoking thoughtful reflection on a broad scope of issues raised by nanotechnology. Ethical concepts, presented in clear and concise lay terms, are used to articulate assumptions and questions that can serve as the basis for common dialogue and problem solving at every level of nanotechnology assessment between any combination of interested parties. Throughout the book, the reader is invited to reflect on the implications of nanotechnology from both personal and professional perspectives.

The book is conceived and presented in three distinct sections. In Part One, *Foundations*, I will provide an overview and conceptual foundation for the rest of the book. In Chapter 1, *Approaching the Nano-Age*, I will briefly introduce the reader to the current nanotechnology landscape and the need to avoid the type of polarizing dialogue that has characterized the debates surrounding genetically modified organisms and stem cell research. Chapter 2, *Ethical Dimensions of Science and Technology*, serves to present key assumptions of the scientific view, describe science and technology as agents of social change, and argue for the role of scientists, engineers, corporations, and regulators as moral agents. In Chapter 3, *Societal Impacts and Perspectives*, I will examine the range of societal impacts that are open to consideration, the important interface of science and human values as evidenced by the Human Genome project, implications of the precautionary principle, and calls for more transparency and public engagement in key aspects of nanotechnology assessment activities. Finally, in Chapters 4, *The Language of Ethics*, and 5, *Method and Process in Ethics*, the reader is introduced to a language, framework, and method for constructing personal and collective reflection on the issues raised by nanotechnology.

The goal of the chapters in Part Two, *Emerging Issues,* is to illustrate key concepts in the proposed reflective framework as applied in the areas of nanomaterials and manufacturing generally (Chapter 6), the military and national security arenas (Chapter 7), sustainability and the environment (Chapter 8), and health and medicine (Chapter 9).

Part Three, *The Framework Applied,* is intended to bring the discussion full circle. In Chapter 10, the reflective framework will be applied in somewhat more depth in a case presentation related to human enhancement technologies. A summary of the pressing questions, the players, the context, the stakes, and a call to action for citizens of the evolving Nano-Age will be proposed in Chapter 11, *The Ethical Agenda for Nanotechnology.* Finally, the book will close with Chapter 12, *Personal Reflections on Nanotechnology and the Moral Imagination.*

About the Author

With 30 years experience in the health care industry, Deb Bennett-Woods EdD, FACME, brings an interesting combination of experience in strategic leadership, organizational development, management theory, and applied ethics to her work in the area of nanotechnology and other emerging technologies.

She holds a B.S. in Psychology and Vocational Rehabilitation, an M.A. in Counseling Psychology, and a doctorate in Educational Leadership from the University of Northern Colorado.

She is an associate professor in the Department of Health Care Ethics at Regis University in Denver, CO, where she also directs the Center for Ethics and Leadership in the Health Professions. She is a Fellow of the Center on Nanotechnology and Society in the Chicago-Kent College of Law at the Illinois Institute of Technology. She is also a board-certified Fellow of the American College of Health Care Executives and holds credentials with the American Health Information Management Association. Prior to her current position, she was the founding Director of the Department of Health Care Ethics and the Director of the Department of Health Services Administration and Management, also at Regis University.

She currently teaches courses with an applied focus in health care ethics across all graduate and undergraduate programs in the Rueckert-Hartman School for Health Professions. From 1993 to 2000, she was the Director of the Department of Health Services Administration and Management at Regis University where she administered programs in Health Information Management, Medical Imaging Management and Health Care Administration, and coordinated the design and launch of the graduate program in Health Services Administration. Prior to 1993, she had 12 years of administrative experience in health care settings.

She was a member of the Task Force on Social and Ethical Implications of Nanotechnology of the Colorado Nanotechnology Initiative and currently serves on the Environmental, Health & Safety Committee of the Colorado Nanotechnology Alliance. She has presented at both national and international conferences on the topic of ethical issues in nanotechnology. Her particular interests lie in the areas of emerging technologies in health care, and the convergence of technologies such as nanotechnology, biotechnology, artificial intelligence, cognitive science, and robotics as they relate to health care and related social policy, as well as the deeper philosophical questions regarding human enhancement and life extension.

She is published on a variety of topics including two proprietary texts commissioned for use by Regis University and The College Network entitled *Ethics and Society* (2002) and *Research in Nursing* (2003).

List of Contributors

Jürgen Altmann
Physicist
Experimentell Physik III
Universität Dortmund
Dortmund, Germany

Rosalyn Berne
Associate Professor of Science, Technology and Society
School of Engineering and Applied Science
University of Virginia
Charlottesville, Virginia

David M. Berube
Professor of Speech Communication
NanoCenter
University of South Carolina
Columbia, South Carolina

Nigel M. de S. Cameron
Associate Dean and Research Professor of Bioethics
Director, Institute on Biotechnology and the Human Future and
the Center on Nanotechnology and Society
Illinois Institute of Technology
President, Center for Policy on Emerging Technologies (CPET)
Washington, D.C.

David Guston
Director, Center for Nanotechnology in Society
Associate Director, Consortium for Science, Policy & Outcomes
Arizona State University
Tempe, Arizona

Jacob Heller
Policy Associate
Foresight Nanotech Institute
Menlo Park, California

George Kimbrell
Staff Attorney
International Center for Technology Assessment (CTA)
Washington, D.C.

Bruce Lewenstein
Professor of Science Communication
Cornell University
Ithaca, New York

Michael Mehta
Professor and Chair, Sociology of Biotechnology Program
University of Saskatchewan
Saskatchewan, Canada

Meyya Meyyappan
Chief Scientist for Exploration
Center for Nanotechnology
NASA Ames Research Center
Silicon Valley, California

Christine Peterson
Founder and Vice President for Public Policy
Foresight Nanotech Institute
Menlo Park, California

Rocky Rawstern
Founder and Consultant
Access
http://access-nanotechnology.com

Daniel Sarewitz
Director, Consortium for Science, Policy & Outcomes
Arizona State University
Tempe, Arizona

Nora Savage
Environmental Engineer
Office of Research and Development
Environmental Protection Agency

Anita Street
Environmental Scientist
Office of Research and Development
Environmental Protection Agency
Washington, D.C.

Chris Toumey
Director, South Carolina Citizens' School of Nanotechnology
NanoCenter
University of South Carolina
Columbia, South Carolina

Gregor Wolbring
Research Scientist
University of Calgary
Alberta, Canada

PART ONE

Foundations

With our thoughts we make the world.

—Buddha

1

Approaching the Nano-Age

1.1 Two Sides of the Same Coin

> In the Land of Mordor where the Shadows lie,
> One Ring to rule them all, One Ring to find them,
> One Ring to bring them all and in the darkness bind them,
> In the Land of Mordor where the Shadows lie.

J.R.R. Tolkien, The Fellowship of the Ring[1]

Despite Tolkien's insistence that his *Lord of the Rings* trilogy should not be read allegorically, the work is often interpreted as a cautionary tale against modernity in general and technology in particular (Schick 2003).[2] The idyllic simplicity of the land of the Hobbits is contrasted with the wasteland of Mordor, laid barren by Sauron's relentless quest for power. The rings are themselves products of technology, forged with great craftsmanship with two intentions. The three elven rings were created with the intent of restoring, healing, preserving, and enhancing Middle Earth. The other rings were created to provide wealth and dominance to those who manage to possess one. The One Ring, capable of forging an abiding and unlimited power to consolidate the rest, can be seen to represent both the obsessive desire for material power and the moral burden that comes with having such power. At one level, and perhaps the level overtly intended by Tolkien, the tale simply cautions against the temptations inherent in placing too much faith and importance in objects of the material world. At a metaphorical level, the power represented by technology is more complex. It can be characterized as having both the positive and healing potential of the elvin rings and the more negative and corruptive power of the other rings—the dualistic nature of all technologies to harbor both benefit and risk.

These two general themes have long been played out in the literary imagination of fantasy and science fiction, and in the technological imagination of popular culture. On one hand, technology is viewed, and rightfully so, as a primary engine of human progress. In many ways, it is the ultimate expression of human intellect, logic, ingenuity, and curiosity. The opening sequence of the original Star Trek series characterizes the mission of the Starship Enterprise in terms that reflect a modern appreciation for human knowledge and technological power: "To explore strange new worlds. To seek out new life and new civilizations. To boldly go where no man has gone before." The original Star Trek series aired in the 1960s, at a time when the race to space was framed as an epic struggle in the Cold War. The crew's weekly adventures captured the optimistic imagination of a technologically advanced future filled with instant communication, teleportation, intelligent computers, intergalactic travel, and an Earth at peace, with the perils of the nuclear age a distant memory.

At the same time, a darker and more menacing theme emerged in later incarnations of the original series. This alternate theme represents the potential for the power of technology to exceed our understanding, capacity, or willingness to use it well, and for the possibility of harmful consequences we did not foresee. For example, consider the Borg, a part-human, part-machine race of creatures characterized by a collective consciousness and total suppression of individualism. The goal of the Collective is the efficient and harmonious existence harbored within a form of technological totalitarianism. Aside from outright destruction of anything in their path, the primary means of achieving this existence is the assimilation of captives into the Collective. Their mantra, "resistance is futile," captures the discomfort many of us feel today with the rapid and uncritical emergence of powerful new technologies that increasingly dominate our lives and seem, at times, to be more in control of us than we are of them.

Regardless of content and view, all technological narratives are morality tales at some level (Schummer 2005).[3] Whether optimistic and forward thinking or pessimistic and cautionary, our judgments about technology reflect deeply held assumptions about who we are, what we value, and the sources of human meaning. Ethics is the language of morality and is clearly evident in even the most objectively rational arguments for or against technological opportunities and imperatives. As such, the common practice of listing ethical issues as a separate realm of societal impact is overly narrow and fundamentally misleading. Ethics is not a checklist of dos and don'ts, nor does it exist solely in the abstract. To the contrary, ethics is an ever-present mediator of human relationships and action. Ethics can be spoken in the technical terms of ethical theory, or in the lay terminology of right and wrong, good and bad, fair or unfair, helpful or harmful, safety or risk. In either case, ethics is the primary framework of meaning in which technology emerges and is assessed.

In the past few years, scholars, researchers, corporate executives, and elected leaders have increasingly discovered that ethics matters. As public trust in our institutions has eroded in the face of scandal, greed, or gross mismanagement, there is growing pressure for transparency and accountability in everything from research to corporate finance and public policy. At the same time, there is a paradoxical apathy and disengagement on the part of many citizens, accompanied by a radical polarization and activism on the part of many others. This somewhat schizophrenic fragmentation of civic engagement complicates social discourse on almost every critical issue we face as a society. Nanotechnology and the techno-cousins it will enable are among those critical issues. To pretend that ethics is an isolated subcategory of societal implications of nanotechnology is a risky and unhelpful oversimplification. The practical and the ethical are two sides of the very same coin. Ignoring the moral dimension of technology is at least as risky as even the most questionable application of technology itself.

Returning to the wisdom inherent in the message of Tolkien, our past, present, and future exist first within our own moral and practical imagination. The overriding message and purpose of this book is to provide an eminently practical means of making apparent our intentions and moral intuitions when considering the complex potential for any disruptive technology, and nanotechnology in particular.

- How will nano- and related technologies improve the human condition now and in the long term?
- How do we weigh the risks and benefits?
- How do we prioritize limited resources in developing nanotechnology's potential?
- How do we minimize disruptions to the workforce and the global economy?
- How do we assess the need for regulation?
- How do we avoid geopolitical conflict over these and other technological advances that threaten the current balance of military, economic, and political power?

These are examples of highly pragmatic questions that warrant serious and well considered answers. Implicit within those answers will be statements of what benefits we value most highly; our definitions of acceptable and unacceptable risks and harms; how we resolve competing duties; where our loyalties lie; how justly benefits will be allocated; and, who will bear the inevitable burdens. In other words...ethical judgments.

Ethics provides a multi-layered lens through which past, present, and future can be examined critically and with both moral and practical intention. When applied poorly, ethics provides incomplete and faulty justifications. However, when applied well, ethics supports the best of the human capacity for critical analysis by enabling both our rational/objective and our more intuitive/subjective natures.

1.2   Setting the Stage

Imagine yourself considering the future at the dawn of the 18th century, just prior to the generally recognized start of the Industrial Revolution. There was little of what we would recognize today as automation. Most products were still crafted by individual tradesmen, on a relatively small scale, and sold in small shops within a local marketplace. Political and economic power was largely concentrated among landholding elite in a primarily agrarian economy. How easy would it have been to predict the far-reaching effects of the rapid succession of technological innovations including the cotton mill, the steam engine, and mechanized production in the factory system?

The Industrial Revolution is credited with a wide array of discoveries and inventions that altered the very fabric of society. There is no doubt that the progress observed during this revolutionary period ultimately contributed greatly to the rise of a middle class, the emergence of modern political and economic systems, higher standards of living, and greater access to material comforts. At the same time, the Industrial Revolution is also credited with the creation of a working class, resulting in a large-scale movement of people to industrialized cities. This migration to the cities in search of jobs was accompanied by the attendant problems of low wages, long working hours in dangerous conditions, brutal exploitation of child labor, overcrowded city tenements and slums, environmental degradation, public health challenges, and other societal effects. Even if one had possessed the benefit of a crystal ball, and could have seen all that would come to pass for the next 200 years or so, how would one have weighed the tremendous benefits of innovation and technological advancement against the very real and serious harms that accompanied it?

The term **revolution**, derived from the Latin *revolvere* meaning to turn over, is commonly applied to a relatively sudden and radical change in a situation. For example, an earlier period, that has come to be portrayed as the Scientific Revolution, introduced radical concepts in fields such as astronomy, cosmology, and physics. This fundamental rethinking of basic scientific assumptions challenged existing understandings of nature, human knowledge, and belief. As these new understandings of "science" emerged, they contributed to sweeping cultural, social, and institutional changes that helped establish the foundations of our modern period (Hatch n.d.).[4]

Another example of the use of the term *revolution* applies to political and military movements that initiate the downfall and replacement of existing governments and institutions. The American Revolution represented a sea change in the paradigm of societal governance and the role of citizens in the social enterprise. More recently, what is sometimes termed the Computer Revolution has dramatically altered everything from basic business models and productivity, to global information access and communication, to how we spend our leisure time. For those born after the 1980s, it seems impossible to imagine life without the full array of information technology at our fingertips.

As with the Industrial Revolution, each of these examples can be examined from the standpoint of both benefits and harms. Each "turning" has disrupted economic, political, cultural, and social understandings and systems—realigning boundaries, redistributing power, and forcing adaptation to untoward consequences. It is the character of revolutions to force us to see and think about the world differently, integrate new knowledge into old paradigms of understanding, and develop new responses in the face of change.

In retrospect, were the disruptions of these revolutions the necessary and unavoidable side effects of human progress, or could they have been anticipated and mitigated in some way? Does our contemporary appreciation for the disruptive effects of technological innovation create a duty to foresee potential risks? Once foreseen, do we have a duty to act to prevent or minimize such risks? Who, if anyone, is morally accountable for defining and then weighing the potential benefits and harms? How will the common good be served and what voices will guide its transformation? Does our modern understanding of social justice obligate us to act in ways that do not further disenfranchise or expand those who are already oppressed? How can we be sure that the best of who we are as a human community emerges on the other side of any revolution, technological or otherwise?

1.3   Nanotechnology: Revolution or Evolution?

Nanoscience and nanotechnology are often referred to in revolutionary terms. Roco and Bainbridge (2001) describe "a revolution in science and technology, based on the recently developed ability to measure, manipulate, and organize matter on the nanoscale."[5] They go on to emphasize the "far reaching impact" and "fundamental scientific advances" that "will lead to dramatic changes in the ways material, devices, and systems are understood and created."[5] Even the budget of the National Nanotechnology Initiative promotes the revolutionary spirit of nanotechnology. The final budget for fiscal year 2007 is entitled, *The National Nanotechnology Initiative: Research and Development Leading to a Revolution in Technology and Industry*.[6]

Milburn (2002) also captures the revolutionary character of nanotechnology by describing its potential for disruption.

> Nanotechnology is disruptive in that it puts pressure on other products or processes to realign themselves around its introduction. More importantly, nanotechnology is transformative in the sense that it has the potential, at least in theory, to transform social relations, labour, international economies, and to affect a range of institutions.[7]

Milburn goes on to suggest that nanotechnology has created a *nanotechnological* means for viewing the world. He claims this altered viewpoint results in nanotechnology actually re-shaping our understanding of nature, which then translates directly into how we approach changes to other frameworks, such as the legal and regulatory or the social and ethical.

Just as the Scientific Revolution challenged our understanding of nature, knowledge, and belief, nanotechnology may radically alter our mastery of the world around us (Keiper 2003).[8] If so, it is also likely to alter how we come to relate to that world in practical terms, as well as culturally, socially, and intellectually. And, just as the person standing at the dawn of the Industrial Revolution would have had a hard time predicting what was to come, we may be equally hard pressed to anticipate and guide the unfolding of the Nano-revolution. In fact, given the pace of discovery and the complexity of today's global society, we may be far more challenged to adapt quickly and well to the potential disruptions nanotechnology may bring.

Others argue against characterizing nanotechnology in revolutionary terms. Life itself, at the molecular level, consists of biological structures and processes operating at the nanoscale. Likewise, nanomaterials have existed for centuries; we simply have not been in a position to control their production and precise effect down to the nanoscale. In fact, nanotechnology itself is not actually new if one considers examples such as the use of copper and silver nanoparticles in paints and glazes by Renaissance artisans to achieve certain colors, especially gold and ruby-red, and iridescence or luster (Padeletti & Fermo 2003).[9]

Along this vein, Keiper (2003)[8] differentiates between "mainstream" nanotechnology, a simple evolution of chemical engineering and materials science, and molecular manufacturing, a more radical and controversial application of the nanoscale. He sees mainstream nanotechnology, which has the potential to yield chemically and mechanically enhanced materials and processes, as more or less unproblematic. On the other hand, molecular manufacturing involves manipulation of materials and processes at the atomic and molecular level by machines that are also constructed on the molecular scale. Examples of potential applications include molecular assemblers, computers, and robots, largely inspired by the biochemical processes of living cells that also occur at the molecular level. Of particular concern to critics is the potential for such machines to be self-replicating. It is molecular manufacturing that is most closely associated with the more radical applications of nanotechnology, such as nano-robots and computers, advances in human or superhuman artificial intelligence, biologically enhanced human performance, and life extension beyond the normal human lifespan. While he acknowledges the potential revolutionary nature of molecular manufacturing, should it even prove technologically feasible, Keiper argues against the present confusion and hype of rolling the two forms together.

Berube (2006)[10] echoes this distinction, describing nanotechnology in the next 5 to 10 years as "not without some risks, but hardly as revolutionary or apocalyptic as those associated with universal [molecular] assemblers." He argues stridently that exaggeration of either the potential benefits or harms of nanotechnology will undermine our ability to assess nanotechnology in a reasoned societal discourse. Whether via public backlash or withdrawal of investor and market support, this lack of public understanding and dialogue may ultimately prevent the realization of nanotechnology's actual potential.

In all likelihood, history will be the judge of nanotechnology's impact and may well do so on the basis of its societal effects rather than its technological accomplishments. If nanoscience and its applications roll out incrementally, at a relatively moderate pace, with largely positive impacts to the national and global economies, workforce, environment, and culture, then it may be viewed in more evolutionary terms. If, however, advances are rapid and disruptive in their effects in the social realm, then the term revolution will more likely apply as we are forced to adapt quickly at many levels. In either case, the results will be judged in moral terms, at the intersection between technology and our values. Ethics is the language in which it will be described and analyzed.

In the remainder of this introductory chapter, I will lay out the key elements of the context in which nanotechnology is unfolding, beginning with a few definitions and a brief overview of the National Nanotechnology Initiative (NNI). I will briefly introduce the scope of societal implications research and discuss three factors that present unique challenges with respect to nano-driven technological innovation.

1.4   Nanotechnology Defined

A technical discussion of nanoscience and technology is beyond the scope of this book and, frankly, redundant for the informed reader. However, for the lay reader, it may be helpful to consider a very brief overview of terms and concepts. The prefix "nano" can be traced to the Greek word *nanos*, meaning dwarf. It is used presently to indicate one billionth. A **nanometer (nm)** is one billionth of a meter, roughly ten times the size of an individual atom. A common comparison places one nanometer at approximately 100 times smaller than the diameter of a single human hair. **Nanoscience** involves the study of materials at the nanoscale. From this point on, terminology becomes less clear, as there are not internationally standardized definitions of many nano-related terms.

Generally speaking, a **nanoparticle** is a particle with the dimension of 100nm or less, the size at or under which novel, size-dependent properties of physics at the nanoscale typically develop. These properties differentiate nanoparticles from the same material in bulk. It should be noted that nanomaterials may be larger than 100nm, e.g., carbon nanotubes, but must have at least one structural feature at the nanoscale. The unique properties of nanoparticles and materials are related to greater surface area per unit volume, which leads to increased chemical reactivity, and size-dependent changes in magnetic, optical, and electrical properties.

With this in mind, there is no standard definition of nanotechnology. Perhaps the most widely recognized definition, which itself has undergone revision, is that of the National Nanotechnology Initiative (NNI website n.d.).[11]

Nanotechnology is the understanding and control of matter at dimensions of roughly 1 to 100 nanometers, where unique phenomena enable novel applications. Encompassing nanoscale science, engineering, and technology, nanotechnology involves imaging, measuring, modeling, and manipulating matter at this length scale.

Nanotechnology research and development, according to the NNI website, is largely concerned with creating improved materials, devices, and systems that exploit nanoscale properties. Ostensibly, this includes the entire continuum between advanced material science and molecular engineering discussed earlier. For purposes of this book, nanotechnology will be considered in this broader sense, recognizing that the nature of societal implications may be quite different depending on how one defines nanotechnology itself. Also, as a matter of clarification, references to nanotechnology can also be assumed to include the underlying nanoscience.

Finally, nanotechnology is viewed largely as an enabling technology and is often paired with biotechnologies, information technology, cognitive science, and robotics in discussions of its societal implications. Therefore, when reference is made to related technologies, it is these commonly cited examples to which I refer.

1.5   Societal Implications and the NNI

Any consideration of the societal ramifications of nanotechnology in the United States must include the context in which nanotechnology is unfolding, beginning with the establishment of the National Nanotechnology Initiative (NNI) in 2001. This federal initiative, coupled with a number of state initiatives, represents the driving force behind nanotechnology research and development at this point. With strong support during the final years of the Clinton administration, efforts as early as 1998 to coordinate programs in nanoscience and development culminated with Public Law 108-153, the 21st Century Nanotechnology Research and Development Act. Signed into law by President George W. Bush in 2003, this act established an infrastructure within the federal government for the coordination and funding of nanoscience and technology research and development.

The current goals of the NNI are stated on its website as follows:

- Maintain a world-class research and development program aimed at realizing the full potential of nanotechnology;

- Facilitate transfer of new technologies into product for economic growth, jobs, and other public benefit;
- Develop educational resources, a skilled workforce, and the supporting infrastructure and tools to advance nanotechnology; and,
- Support responsible development of nanotechnology (NNI n.d.).[12]

With these goals in mind, seven individual Program Component Areas (PCAs) are delineated. These areas represent the major categories among which strategic investment of the NNI budget is divided. They include:

- Fundamental nanoscale phenomena and processes
- Nanomaterials
- Nanoscale devices and systems
- Instrumentation research, metrology, and standards for nanotechnology
- Nanomanufacturing
- Major research facilities and instrumentation acquisition
- Societal dimensions (NNI n.d.).[13]

Situated within the framework of the National Science and Technology Council, the initial budget of the NNI was $3.7 billion over four years. A report by the President's Council of Advisors on Science and Technology (2005),[14] entitled *The National Nanotechnology Initiative at Five Years: Assessment and Recommendations of the National Nanotechnology Advisory Panel*, provided an overview of progress in the NNI. Approximately 8% of the total NNI budget for 2006, or $82 million, was devoted to the category of societal implications. The Council concluded that this current level of funding was appropriate to the task. It is useful, however, to see how this money is allocated among the various categories of societal implications. Approximately $38.5 million was earmarked for programs working on environmental, health, and safety research and development. The remaining $44.6 million was split between "education" and "ethical, legal, and other societal issues"—a rather wide scope of investigation if the predictions of NNI supporters regarding the revolutionary nature of nanotechnology are accurate.

Activities of the NNI currently encompass the participation of 26 federal agencies with a fiscal year 2008 budget projection of $1.5 billion. Of this, $97.5 million is earmarked for societal dimensions, less than 7% of the total budget. Of the $97.5 million, $58.6 million is dedicated to environmental, health, and safety research and development, leaving $38.9 million for other aspects of societal dimensions (NNI n.d.).[15]

1.6   Societal Dimensions of Nanotechnology

So, what specifically are the societal dimensions included in the Societal Dimensions PCS? The current dialogue on the societal implications of nanotechnology has deep roots in a workshop sponsored by the National Science and Technology Council's (NSTC) Committee on Technology (CT), via its Subcommittee on Nanoscale Science, Engineering and Technology (NSET), in September of 2000. Invited participants represented a cross section of academia, government, and the private sector. The purpose of the workshop was to review current studies on the societal implications of nanotechnology. The range of implications was defined broadly to include educational, technological, economic, medical, environmental, ethical, legal, and other impacts (Roco and Bainbridge 2001).[5] A second goal was to identify methods for conducting investigation and assessment of social implications research in the future. The third goal, and most pertinent for our purposes here, was "to propose a vision for accomplishing nanotechnology's promise while minimizing undesirable consequences."

The workshop yielded a report entitled *Societal Implications of Nanoscience and Nanotechnology* that has been influential in framing the dialogue surrounding societal implications of nanotechnology as well as funding priorities. The report is a compilation of expert statements on what we might expect both generally and specifically. It characterizes nanoscience and nanotechnolgy as a scientific and technological revolution with this being "a rare opportunity to integrate the societal studies and dialogues from the very beginning and to include societal studies as a core part of the NNI investment strategy" (Roco and Bainbridge 2001).[5]

In reading the report, it is easy to understand how nanotechnology has come to be viewed as revolutionary in nature. In and of itself, the breadth of industries affected signals the potential for sweeping change in the national and global marketplace with obvious implications for economic competitiveness and stability, workforce preparation, and an array of policy considerations. A brief overview of the goals for nanotechnology, as outlined by the workshop participants, includes: high performance materials and new manufacturing processes; enhanced electronics; improved health care including advances in pharmaceuticals; sustainability impacts related to green technologies, energy efficiency, water purification, and agriculture; expanded capabilities in space exploration; military and defense applications for improved national security; and continued economic competitiveness.

As might be expected, the report conveyed only a mildly precautionary note insofar as it called for making social and economic research studies a high priority. The report was generally optimistic about nanotechnology's positive potential to "fundamentally transform science, technology, and society" in the next 10 to 20 years. This widely published report and its attendant recommendations have done much to define the direction of societal implications research including funding priorities and the tone of policy.

A similar workshop was held in December of 2003. Again sponsored by NSET, this workshop gathered a range of experts to comment and offer recommendations on ten "transformative themes" including:

- Productivity and Equity
- Future Economic Scenarios
- The Quality of Life
- Future Social Scenarios
- Converging Technologies
- National Security and Space Exploration
- Ethics, Governance, Risk, and Uncertainty
- Public Policy, Legal, and International Aspects
- Interaction with the Public
- Education and Human Resource Development

Many of the findings of the report will be revisited in later chapters; however, this list of themes provides an idea of the scope of issues falling under the general category of societal dimensions (Roco and Bainbridge 2005).[16] Major recommendations focused on risk assessment, public engagement, and education. Of particular note, the vision of a model of responsible technological development included the development of advanced ethical and societal implications, educational resources for K-12 and college courses, as well as continuing education for companies, scientists, and engineers.

It is important to note that ethical dimensions are never defined in these reports, or most anywhere you find the term used. "Ethical" is routinely tacked on at the beginning or end of "societal"; however, what exactly differentiates ethics from societal dimensions is rarely made apparent. The consistent separation of ethical and societal implications in these reports may encourage the image of ethics as an abstract ideal rather than a practical tool for analysis.

1.7 Factoring in Pace, Complexity, and Uncertainty

At this point, it should be clear that nanoscience and technology may have broad effects across many, if not all, industry sectors. In addition, although the NNI is clearly a driving force for nanotechnology in this country, the context of nanotechnology is global in scope. Major investments are also being made in Europe and Asia, with major breakthroughs being viewed as keys to future economic prosperity, and perhaps even global dominance (Berube 2006).[10] Given the magnitude of political and economic investment and both the effort and the potential breadth of impact, at least three other elements are necessary for the reader to critically appreciate the context in which the moral dimensions of nanotechnology can best be considered: pace, complexity, and uncertainty.

Nanoscience is generally acknowledged to be in its infancy (Lane and Kahlil 2005).[17] Most of the many projected applications remain speculative at best. So, how does one assess the potential implications of a technology that mostly doesn't yet exist? And, perhaps most importantly, why should we try?

Assessing the societal outcomes of a major technological innovation in just one single industry involves consideration of a wide array of potential impacts. Consider the cell phone, an innovation that shook the entire telecommunications industry to the core. Cellular technology led to fundamental changes in infrastructure, displacement of prior industry leaders, and an entirely new retail market for everything from the phones themselves to designer ring tones. Rapid growth of the market led to significant disruptions in the workforce as companies emerged, merged, right-sized, downsized, or simply disappeared. However, the effects extend well beyond the business environment as cell phone ownership has skyrocketed in the last 10 years. Concerns range from potential health risks, such as brain tumors in high usage users, to distracted drivers, to airline safety, to tracking terrorist communications. Cell phones have added to the 24/7 feel of the ever-lengthening workday as we are expected to be more and more readily available to our employers. A similar theme emerges as we consider the same 24/7 availability to friends and family who inevitably ask, "Why didn't you answer your phone when I called 10 minutes ago?" Even late adopters who cling to their land lines are barraged by obnoxious ring tones in the theater and loud, one-sided conversations in restaurants, grocery lines, and every other conceivable public location, not to mention the incredulous question, "You don't own a cell phone?!"

On the positive side, working parents have a much easier time keeping track of children; first responders are more readily called to the scene of an accident or crime; and time sensitive communications can be scheduled even in the midst of a busy travel day regardless of location. On a global level, cell phones have become a technological equalizer as millions of citizens of developing countries, for whom traditional phone service was simply not available, now have unprecedented access to family, friends, and the outside world. And, of particular importance, is the extraordinarily rapid dissemination of cell phones to something approaching global market saturation in just over 10 years. Compare this with traditional land lines that, as measured by the census, took nearly 100 years to reach the 90th percentile of households in the United States.

Now take the magnitude of change in one industry and extrapolate it to virtually every industry sector, all more or less simultaneously, as advances in nanotechnology in one industry cascade through other industries as well. Clearly, the escalating pace of discovery and the speed with which today's emerging technologies can be disseminated globally is unprecedented in human history. The concept of the "historical exponential view," popularized by the writings of Ray Kurzweil (2005, 1999),[18, 19] combines the historical observations that human progress has tended to be exponential rather than uniformly linear, with the prediction of a point at which exponential growth becomes "explosive and profoundly transformative." In other words, we have to anticipate that the rate of progress we experience 10 years from now will be greater than the rate we experience today, just as today's rate of progress is greater than the rate of 10 years ago. In addition, there may be a point at which the very idea of prediction becomes irrelevant.

Despite strong evidence of this exponential trajectory, we also have a tendency to underestimate progress and the pace of change. Take, for example, the Human Genome Project, launched in 1990 with a projected 15 year time frame that many skeptics felt was itself overly optimistic. Yet, the project to map the human genome was completed in 2003, under budget and well ahead of schedule, not to mention even further ahead of any reasonable societal consensus on how to approach the practical and moral dilemmas this powerful technological advance represents. This particular example brings to mind the Parable of the Boiled Frog, popularized by management consultant Peter Senge (1990).[20] A frog placed in hot water will immediately attempt to escape; while a frog placed in cool, but slowing heating water will adapt until it cannot escape. The controversies of genetics were not hard to anticipate; however, serious societal dialogue was largely rejected on the grounds it was too early in the process to formulate a meaningful response. At present, major advances in genetic testing, diagnosis, identification of genetic markers, and genetic manipulation have already arrived; commercialization is well underway; and, the field is continuing to advance rapidly as disparate streams of research and development converge. Yet, as with our friend the frog on a slow boil, we have done relatively little to address questions of genetic privacy and discrimination, patent and ownership issues, prenatal screening and genetic selection, access to genetic services by the poor or underinsured, the real and looming possibility of precise genetic manipulation for eugenic purposes, and other related concerns. In fact, nanotechnology provides intriguing possibilities as an enabler for genetic engineering and repair—an area of genetic technology with obvious potential for both highly beneficial and morally questionable applications and abuse.

So, pace presents a challenge in and of itself. The more rapidly knowledge and its applications emerge, the less time we have to reflect and consider how best to use them and what, if anything, to limit. Predictions on the pace of NT research and development vary widely and for good reason. Nanotechnology is inherently multi-disciplinary, leading to an unprecedented level of collaboration between formerly separate research disciplines. This in and of itself complicates projections, although we can perhaps assume the increased efficiency of information sharing and creative synergy is most likely to expedite the process of discovery. With the general difficulty of prediction in mind, most predictions fall into short-, medium-, and long-term time frames with long term being only about 20 years. Inherent in this linear trajectory are assumptions about which aspects of nanoscience and technology will arrive in the short-term to enable applications in the medium-term and so on. Any deviation has the potential to speed up or slow down any number of knowledge sensitive paths along the way. Similarly, many of the medium- and long-term applications of nanotechnology will be dependent on the convergence of other emerging technologies, such as biotechnology and cognitive science, adding yet another level of unpredictability to the mix.

At the same time, technology is speeding up and becoming more technically complex; the social order itself is becoming inherently more complex. Globalization, economic insecurity, political instability, escalating military conflicts, and other avenues of societal change are critical lenses through which technological priorities and advances are viewed. Global warming, the "war on terror," and pandemic flu are examples of societal issues that can easily dominate everything from research funding, to regulatory discussions, to public opinion polls, and the editorial page. A seminal event, such as the attacks of September 11, 2001, can alter the focus of societal priorities in the space of a heartbeat, with the effects lingering for years.

In that same vein, more mundane factors, such as investment, swings in the financial markets, consumer demand, government subsidy of outmoded industry sectors, patent infringement and other legal challenges, corporate failures, political agendas, grassroots activism, and random natural disasters can individually, and in combination, alter the trajectory and timeline of research and development activities (Mulhall 2002).[21] As noted by Mulhall, "Looking at such divergent variables, we see that it's impossible to consider only the scientific side when we try to forecast what science means for us in the 21st century. While progress in technology may seem inevitable, it's by no means guaranteed."[21]

These highly variable determinants of pace lead us to the second critical factor, complexity. The term complexity defies precise definition here, but is intended to generally refer to the emerging insights of general systems theory and the science of complexity. Complexity refers to a condition somewhere between order and disorder, in which multiple parts are simultaneously behaving independently and in ways that are closely interconnected and dependent. Complex adaptive systems (CAS) are collections of component parts, or agents, that are capable of adapting and evolving based on their interactions with each other and with the environment. Such systems are characterized by high levels of ambiguity, dynamic capability, and resistance to hierarchical control.

Society itself is a large, complex adaptive system composed of many smaller systems and individual agents. These various parts are highly interdependent, often interacting in unexpected and non-linear ways. This high level of interconnection means that changes in one part of the system may cause unintended or unanticipated changes in other parts of the system, that may, in turn, create conditions in which entirely new and novel events or characteristics can emerge. The emergent properties of complex systems contribute to their ability to spontaneously self-organize and adapt in response to changes within the system or new conditions in the environment.

We have increasingly come to recognize that our historical assumptions of an essentially linear and reducible universe are incomplete and inadequate to understanding the dynamics of everything from basic physics, to biological systems, to human society. Yet, these assumptions still drive much of our analysis and problem solving at the societal level. The condition of complexity within a system makes it necessary to view the system in terms of patterns of relationships rather than narrow cause and effect responses. Effective deliberation on emerging technologies must include an acknowledgement that complex systems are inherently hard to model, predict, and control. Simplistic assessments based on narrowly targeted risks or projected economic benefits will inevitably fall short. Attempts to draw clear boundaries around categories of societal impacts or to isolate stakeholder groups as a method of analysis will likewise prove unsatisfactory.

Finally, the combination of rapid pace and inherent complexity lead to the third factor of uncertainty. Simply put, uncertainty reminds us that we don't know what we don't know. In strategic planning, the concept of uncertainty encompasses a range of factors that are unrecognized, unpredictable, uncontrollable, or perhaps even, unknowable. Courtney (2002) terms the uncertainty that still exists after the best possible analysis as residual uncertainty.[22] He proposes four levels of residual uncertainty. Level One situations contain a great deal of predictability among many potential factors with a single likely outcome being identifiable. Levels Two and Three are characterized as having a range of possible outcomes with varying degrees of reliability in their prediction. Level Four uncertainly is the level most ambiguous and resistant to analysis. There is a wide range of variables, known and unknown, to be factored into an analysis, and very little information of solid predictive value. This fourth level is the level of uncertainty common at the boundary of major technological, economic, or social discontinuities. It is also common at the point at which new markets are just beginning to emerge or when extended time frames are required to evaluate strategies. The current state of nanoscience and nanotechnology research and development falls easily within this description.

Uncertainty clearly complicates our ability to apply purely objective criteria to the decision-making process. If given too much weight in the process, high levels of uncertainty can become paralyzing, as risk assessments are elevated based on the absence of predictive information. On the other hand, uncertainty can also act in the opposite direction. In complex systems, if we are overly narrow in our impact modeling, or if we truly don't know what we don't know, then we can be tempted to move ahead too quickly or confidently and we can fail to recognize risks, harms, negative trends, or additional uncertainties as they emerge.

COMMENTARY

Nanotechnology: Making Real Choices

Jacob Heller and Christine Peterson

Nanotechnology has immense potential. Within the next few years, nanotechnology is expected to be used in important and diverse applications, from localized cancer treatments, to powerful new solar technologies, to super strong yet lightweight materials. In the longer term, productive nanosystems, or the capability to build macroscale objects from the molecular level up, should revolutionize the way our products are produced, making it possible to fabricate any object with atomic precision. To advocates and observers of nanotechnology, the many potential future benefits of nanotechnology are clear. It is equally clear that such powerful technologies might also have negative consequences, such as endangering our health, our environment, or triggering a nanoweapons arms race. Our goal should be to guide nanotechnology's development in order to maximize its potential benefits and minimize its downsides.

It is crucial that concerned constituents begin to consider the implications of certain aspects of nanotechnology and participate in informed debate now, since today's choices will necessarily determine tomorrow's technologies. If we were to judge, for instance, that some nanotechnologies are potentially so environmentally dangerous that all development should be halted immediately, nanotech development would slow dramatically. On the other hand, if we were to judge that the environmental implications of nanotechnology were negligible, its development would forge ahead unfettered and unregulated, whatever the risks may be. Choosing not to make judgments on matters like this is still a choice: it is a decision that the future of this important technology is left to a combination of special-interest political motives and markets. For those who believe that these forces alone may not make the best choices for us as a society (and history provides ample evidence that they will not), it is important that we start considering the often difficult and complex matters of policy for ourselves.

Three broad policy areas of present and ongoing political and ethical debates will bear greatly on the future of nanotechnology: innovation, regulation, and societal implications. Innovation policies are the ways that governments can help ensure specific nanotechnologies are developed, and determine which applications of nanotech are encouraged (and by implication, which are not). Such policies include patent law and government financing of nanotech research and development. Regulation policy concerns the areas of nanotech development that should arguably be controlled, including any potentially harmful nanoparticles, nanomedical alterations made on those unable to choose for themselves (children, for example), and the export of nanotechnologies that could be used as weapons. Finally, there is current policy discussion on the possible implications of nanotechnology, including its effects on poverty, inequality, and privacy. How independent organizations, governments, and their citizens all over the world approach financing of nanotechnology, regulate its potential risks, and deal with its implications will shape the future of nanotech and its impact on humanity.

Most of these issues will likely be covered in other commentaries of this volume in great detail. What is important to keep in mind, however, is that we no longer have the luxury that time once afforded us; foresight, with all its benefits, is a commodity we are quickly running short of. Nanotechnology, which may have seemed far in the future only a decade ago, is already making an appearance in nontrivial ways. Its increasingly rapid development, evidenced by its growing coverage in scientific journals, the rising levels of public and private financing going towards nanotech ventures, and, some would say more ominously, the ballooning number of sometimes overlapping patents being filed with nanotechnological innovations at their core, proves it is no longer something of the future, but of the present. The time to inform, prepare, and decide for yourself is today.

Jacob Heller, a Policy Associate with the Foresight Nanotech Institute, received his Honors B.A. in Political Studies and Economics from Pitzer College in December 2006, and will be attending Stanford Law School in the fall of 2007. Deeply interested in the intersection between law, technology, and society, he founded an organization, A Computer in Every Home, that provides free laptops and computer training to disadvantaged students.

Christine Peterson is Founder and Vice President for Public Policy of the Foresight Nanotech Institute, the leading nanotech public interest group. She serves on the Advisory Board of the International Council on Nanotechnology, the Editorial Advisory Board of NASA's Nanotech Briefs, and on California's Blue Ribbon Task Force on Nanotechnology. She is a co-author of *Unbounding the Future: The Nanotechnology Revolution* (William Morrow and Company 1991) and *Leaping the Abyss: Putting Group Genius to Work* (knOwhere Press 1997).

Uncertainty is not a comfortable condition for most human beings. It gives rise to resistance on the part of people who are not committed to change in the first place. It can generate defensiveness on the part of proponents of change. Both reactions can result in combative and adversarial approaches to planning and response that lack the necessary collaborative and creative energies to respond quickly in complex, fast-paced environments. On the other hand, both resistance and defensiveness can be channeled productively, if well facilitated, on the basis of common goals and meanings. In the best sense then, uncertainty can operate to keep us humble and more responsive to subtle signs of the need to slow down or change course.

1.8   Citizenship in the Nano-Age

> Nanotechnology is emerging into a global system in which a new sense of human responsibility is evolving.
>
> **—Michael Mehta and Geoffrey Hunt[23]**

Rapid pace, increasing complexity, and high levels of uncertainty will characterize any meaningful consideration of nanotechnology and its convergent technological partners. All indications are that the pace of technological discovery and innovation will continue to accelerate, leaving shorter and shorter time frames in which to anticipate societal impacts in an ever more complex and uncertain human system. However daunting this combination of challenges may sound, there is reason for genuine optimism and excitement. For all its unknowns, nanotechnology also represents an unprecedented opportunity to adapt to the challenges by forcing us to create new, more innovative, and effective methods of forecasting, prioritizing, planning, assessing, communicating, collaborating, educating, and regulating any technological imperative. Whether we rise to this challenge will depend on the extent to which scientists, policy makers, regulators, corporate interests, targeted interest groups, and the general public accept the mantle of responsible, informed, and collaborative citizenship. The impact of nanotechnology, evolutionary or revolutionary, will largely be as positively or negatively disruptive as we allow it to be. If and how the stated commitment to examine societal implications rolls out in practical terms is where the rubber hits the road. Using ethics as a framework for navigating these relatively uncharted waters of technological emergence is one proverbial step in the right direction.

1.9   In Summary

We have covered a great deal of territory in this chapter in order to establish the context within which ethics can be used as a tool for technological assessment. The dual impact of major technological advances was illustrated in both literary and historical examples. A very basic introduction to nanoscience and nanotechnology was provided along with an explanation of the role and focus of the National Nanotechnology Initiative. Finally, we began considering the challenges inherent in assessing social impacts of nano- and related technologies, with particular emphasis on the factors of pace, complexity, and uncertainty. In the next chapter, we will explore the relationship between ethics and science and the factors that lead to a natural tension between scientific discovery and its attendant impacts on society.

1.10   Questions for Thought

1. A now famous quotation by Ken Olson, president and founder of Digital Equipment Corporation, in 1977 contends, "There is no reason anyone would want a computer in their home." Brainstorm a list of the impacts of the personal computer on U.S. society in the years since its introduction. If done well, the list will be long.

    Does the list include what might be considered positive and negative impacts? For example, does your list include efficient access to useful information along with a lot of inefficient sorting through irrelevant information and misinformation? What about the convenience of internet shopping along with the rise of internet-based consumer fraud and identity theft? How many of these effects could have been accurately predicted at the time? How might we have anticipated the negative effects and responded more effectively to their prevention or control?

2. Select one of the "transformative themes" identified in the 2003 NSET workshop.

   - Productivity and Equity
   - Future Economic Scenarios
   - The Quality of Life
   - Future Social Scenarios
   - Converging Technologies
   - National Security and Space Exploration

- Ethics, Governance, Risk, and Uncertainty
- Public Policy, Legal, and International Aspects
- Interaction with the Public
- Education and Human Resource Development

Brainstorm possible positive and negative impacts of nanotechnology within the chosen theme. What ethical issues are likely to be raised? How might the dynamics of pace, complexity, and uncertainty affect our ability to predict and respond to potential issues raised within this theme?

2

Ethical Dimensions of Science and Technology

2.1 Ethics at the Intersection of Science, Business, and Governance

> Science cannot resolve moral conflicts, but it can help to more accurately frame the debates about those conflicts.
>
> —**Heinz Pagels**

J.R. Oppenheimer was the theoretical physicist who directed the Manhattan Project during World War II. Sometimes referred to as the "father of the atomic bomb," Oppenheimer reflected on the relationship between science and the larger society in an address to the American Philosophical Society in 1946:

> What is not contingent is that we have made a thing, a most terrible weapon, that has altered abruptly and profoundly the nature of the world. We have made a thing that, by all the standards of the world we grew up in, is an evil thing. And by so doing, by our participation in making it possible to make these things, we have raised again the question of whether science is good for man, of whether it is good to learn about the world, to try to understand it, to try to control it, to help give to the world of men increased insight, increased power. Because we are scientists, we must say an unalterable *yes* to these questions; it is our faith and our commitment, seldom made explicit, even more seldom challenged, that knowledge is a good in itself, knowledge and such power as must come with it.[1]

This core assumption, that knowledge is a good in and of itself, has long served to legitimize science as a good, in and of itself, and to inform the manner in which science then relates to the rest of society. No matter how terrible the effects of the atomic bomb were, regardless of our retrospective questioning of the necessity of their use, and independent of the admittedly "evil" nature they embody, the inherent good and the value of the underlying science is not in question. Likewise, business has generally adopted the stance that science leads to technological innovation, lending itself to new and better products, enhanced competitive edge, expanded markets, and higher profits. In support of both science and business, government has adopted a similar stance that extends the benefits of science and technology to general economic prosperity as well as global dominance and national security.

In most ways, nanoscience and nanotechnology are not different from their predecessors. Any major advance in science or technology is revolutionary in its own way and time. It is routinely predicted that the potential effects of nanotechnology will be far-reaching and perhaps unprecedented in the history of technology. These effects may also emerge faster and in a more complex and uncertain environment than previous technological advances. For these and other reasons, there is value in isolating nanotechnology as a special case when considering its moral implications. However, there is also value in considering nanotechnology within the context of the more general view of science and technology.

Based on their research regarding public attitudes toward the governance of science, Gaskell and his colleagues (2005) argue against the status quo of science policy. Their grounds are that the status quo increasingly risks alienating a key constituency, consisting of a moderate leaning minority of the public, that express interest in being more involved and more attentive to who questions about the type of society we want and the moral and ethical considerations of science and technology in achieving it. These authors conclude:

> These are questions about ethics and social values; science alone cannot answer them. The public expect and want science and technology to solve problems, but they also want a say in deciding which problems are worth solving. This is not a matter of attracting public support for an agenda already established by science and scientists, but rather of seeing the public as participants in science policy with whom a shared vision of socially viable science and technological innovation can be achieved.[2]

In this chapter, I want to explore the core assumptions of a technological and scientific view and some of the moral conflicts inherent in this view as it relates to scientists and engineers, as well as how it spills out into the realms of governance and business. I will also revisit some of the specific challenges of pace, complexity, and uncertainty.

2.2   Assumptions of Science and Technology

> Science is the tool of the Western mind and with it more doors can be opened than with bare hands. It is part and parcel of our knowledge and obscures our insight only when it holds that the understanding given by it is the only kind there is.
>
> —Carl Jung

The history and philosophy of science and technology is a complex area beyond the scope of this book. However, an appreciation for the basic, under-lying assumptions of science and technology is important to societal considerations. For this reason, we turn briefly to an overview of science and technology as they are commonly perceived, conducted, and valued.

The Scientific Revolution, predominantly associated with the 16th and 17th centuries, is popularly credited with the origins of our modern conceptions of science and the scientific method. The work of scientists such as Copernicus, Kepler, and Newton challenged dominant views of nature and the universe and initiated a literal revolution in the method of science and its underlying philosophical assumptions. While science is just one among many ways we "know" the world, it has assumed and maintained a particularly privileged position as perhaps the most consistent and widely accepted source and method of knowledge generation.

The first major building block of modern science to emerge from the Scientific Revolution was the adoption of a mechanistic view of nature. In this view, reality is reducible to a series of physical laws that can be discovered through a combination of intuitive insight, logic, observation, and measurement. These laws are revealed through observable patterns that are consistent over time and place, constituting a scientific truth that can be applied in all similar circumstances. In other words, by the observation and measurement of a thing or a process in its essential parts or sub-processes, we can explain and predict any phenomenon in the natural world.

Embedded within this general worldview are several important foundational assumptions:

- There is one objective reality (ontological assumption)
- The scientist is independent from that which is being studied (epistemological assumption)
- Scientific knowledge is value-free and unbiased (axiological assumption) (Creswell 1994)[3]

These assumptions regarding an objective reality, free from bias and value judgments, and that is discoverable in ways that are independent of the investigator, lead naturally to a set of methodological assumptions. It is assumed that scientific knowledge can only be generated through systematic experimentation in which hypotheses and theories are empirically tested against observations in the natural world. The evidence collected via customary methods of experimental observation and measurement must be rigorously examined against commonly accepted standards of accuracy, reliability, and validity. Once determined to be valid, knowledge is applied toward the methodological ends of prediction, explanation, and understanding.[3]

Key to our purpose in this book is the emphasis on the implicit assumption that an unbiased rationality of science is superior to the irrational and subjective lens of human values, moral tenets, and ethical principles. The scientific view largely rejects, or assigns secondary status to ways of knowing that are not based on a foundation of facts, as defined by science. Values are mere human constructions while the facts of science exist independent of human control or interpretation. This contention that scientific knowledge is value-free has left science fairly insulated from meaningful moral analysis outside of a few mostly non-scientist philosophers in the halls of academe. Science is seen to be cumulative in nature so that, while it may require an imaginative leap to move a scientific theory to the next level, such imagination is necessarily bounded by what is already known to be "true" within the scientific paradigm. Principles and theories are naturally discarded or revised as new and more complete understandings are achieved, but always in terms of a progressively linear trajectory of scientific truth. Questions of value are rarely posed; and, when they are posed, they are answered in the vernacular of science, used by Oppenheimer earlier in the chapter, as an unqualified social good (Chadwick 2005).[4]

Positivism is a term used, in part, to describe a theory of knowledge that relies exclusively on a limited application of scientific method for the production of knowledge and the notion of scientific progress. Positivistic beliefs have played a powerful role in the modern view of science and how scientific knowledge is valued in the larger culture. For example, the belief that science is the only legitimate form of knowledge largely displaces religion and philosophy as valid sources of knowledge. Because science is self-justifying and self-validating, there is no need for the scrutiny of other less verifiable forms of human knowledge (Bentz and Shapiro 1998).[5]

Positivism's view of a single, objective reality implies that all scientific disciplines are valid only to the extent they conform to the methodological assumptions of the natural sciences. This has resulted in the large scale effort by the various human and social sciences to establish empirically consistent methods that allow them to contribute to the eventual construction of a fundamental unified science—if you will, a theory of everything.[5,6] Furthermore, in the positivist view, such movement toward an all encompassing model of science is directly credited with being the primary driver of human and social progress. To the extent that science provides us with the ability to predict and control reality, it also becomes the means for achieving new and higher levels of humanity's individual and social potentials.

Critics of this traditional, and somewhat narrow, view of science as the sole arbiter of truth, and there are many, challenge all of these assumptions at some level even in the natural sciences (Bentz and Shapiro 1998; Harding 2002; Allen, Tainter, Pires, and Hoekstra 2001).[5,6,7] For example, many Western philosophical traditions, from Aristotle to Kant, have insisted on the clear necessity of rationality in moral analysis and justification. In addition, the very argument that science is the source of human progress is subjective and value laden itself, revealing an overt and unsubstantiated bias in the core assumptions of positivism.

The technoscientific view, embodied in the field of science and technology studies (STS), contends that science, technology, and culture are all intimately interdependent. This interdependence is perhaps best illustrated in co-productionist models of knowledge that emphasize the complex and evolutionary nature of the interface between the scientific and social domains of knowledge. The concept of co-production of knowledge is highly critical of what is considered an artificial separation of the objective, factual, reason-based domains of science and policy, and the more subjective, emotional, value-laden domains of culture and politics (Jasanoff 2004).[8] In other words, knowledge generated by science is subject to a variety of social influences just as other forms of social knowledge are influenced by science. Specifically, Jasanoff argues that co-productionist approaches reveal, "unsuspected dimensions of ethics, values, lawfulness, and power within the epistemic, material, and social formations that constitute science and technology."[8]

Despite such recognition of the role and importance of normative concerns in STS discussions, specific moral assumptions and foundations are still not often articulated. For example, in considering the policy discussions related to genetically modified organisms (GMOs), Wynne (2001) points to a lack of acknowledgement of human values and ethical commitments that he believes stems from an artificial divide between categories of risk and ethics. He further charges that by narrowly constructing ethical issues as either scientifically reducible in terms of risk, or as subjective matters of individual choice, science and its institutions have actually escaped the need for a deeper and better informed level of self reflection on their own strengths and limitations.[9]

One more assumption is worth noting here. This involves the perceived difference between basic or pure science and applied science. While the term basic science refers generally to the systematic collection of knowledge through the scientific method, applied science is the application of that knowledge to meeting specific human needs. This basic distinction between the knowledge generated and its ultimate application, generally in the form of technology, further insulates the scientist from the social implications of his or her work.[5] This is evidenced somewhat by the fact that various codes of ethics for scientists focus almost solely on the responsibility to generate new knowledge and avoid any actions that undermine the credibility of either the knowledge generated or the reputation of scientists in general. Engineers, on the other hand, are charged with issues such as public safety and managing the risk associated with new technologies.[10]

This brings us to the overriding point of this section. Regardless of the complex models of interaction between the social and the scientific that are being generated in academic circles, a fundamentally positivist viewpoint still dominates the understanding of science by most scientists, policy makers, and the general public. According to Benz and Shapiro, "positivism, explicitly or implicitly, is at the core of the modern worldview of scientific, technological, bureaucratic, commercial civilization."[5] The resulting insulation of scientists, engineers, policymakers, and business leaders from critical reflection on the "good" of science, makes any effort to initiate serious ethical examination of any particular thread or goal of nanoscience and nanotechnology more difficult than it perhaps needs to be. Naturally, many scientists and engineers do reflect deeply on the implications of their work and the nature of their power and influence, as do many policymakers and business leaders. Unfortunately, even this can take a reactive and defensive turn if the core assumptions of the inherent good of science are questioned too aggressively or in non-productive ways (Levitt and Gross 1996).[11]

At some point, we must be able to recognize that science is value-laden without throwing the proverbial baby out with the bath water. Rationality, objectivity, and other standards of science are also objects of value. What we choose to investigate, what resources we invest, and our intentions for how the knowledge is to be used are driven by values. By acknowledging that science is subject to the same basic elements of social construction as other forms of knowledge, we can begin to build the necessary bridges needed to incorporate practical and moral sensitivities into the process and culture of science more explicitly, and in a way that likewise increases those same sensitivities in the larger society. Given that societal considerations have been explicitly targeted for consideration by the NNI, nanoscience and nanotechnology offer an intriguing testing ground for an expanded awareness.

2.3   Science and Technology as Agents of Social Change

> Western society has accepted as unquestionable a technological impera-
> tive that is quite as arbitrary as the most primitive taboo: not merely the
> duty to foster invention and constantly to create technological novelties,
> but equally the duty to surrender to these novelties unconditionally, just
> because they are offered, without respect to their human consequences.
>
> **—Lewis Mumford**

That science and technology are agents of social change seems imminently apparent. One needs simply to look around to see the gross impact science and technology have on our daily lives, particularly those of us in technologically advanced societies. Science and technology happen largely without the consent of the majority of those likely to be affected. Relatively few people are fully aware of the sheer magnitude of the scientific enterprise or its impact on the economic, political, and social spheres of society. Science itself generally resides below the radar screen of the general public with the exception of interesting television on the Discovery channel, and the occasional attention-grabbing headline about global warming or stem cell research. Yet, it profoundly affects every corner of our lives. Technology, on the other hand, is so ubiquitous that we rarely give it much thought beyond the commercial hype of the newest gadget.

Jonas (2004/1979) proposed four traits of modern technology.[12] The first trait is that technology resists both equilibrium and saturation, which means that technology tends to continue developing and expanding its reach. A second trait is that scientific discoveries and technological innovation spread quickly through the "technological world community," via the technological complex and in response to the pressures of economic and other competition. Third, the ends of technology do not follow a linear pattern in that new technologies generate new, previously unconceived ends beyond those initially sought. It does so simply because it can. Thus, the objectives of technology expand as feasibility expands and in ways we may well not have predicted or even imagined. Finally, progress is an inevitable outcome, with the term "progress" narrowly describing the tendency of the next technological advance to be technically superior in some way.[12]

When considering science and technology as change agents, it is worth mentioning the compulsory nature of technology alluded to by Mumford at the beginning of this section. Each of the previously mentioned traits operates at some level to enhance the compulsory nature of technology adoption. The fact that the natural tendency of technology is to continue to develop and become more available increases the likelihood that we will become aware of it and will be able to acquire it. The dynamic of competition pushes technologies more and more quickly, and more aggressively into the marketplace. The tendency for one technology to generate new and novel technologies means there are even more technologies available from which to choose. Technological progress tempts or forces us to "upgrade" to newer and better technologies as they become available. However, we don't always consider the full range of the impact.

Take, for example, safety features on cars and trucks. Initially, seat belts were optional features that enhanced the market appeal of a vehicle to buyers who could afford the additional expense. Eventually, they became standard, then required, and now seatbelt laws make their presence and use compulsory. Air bags, a novel technology developed in response to some of the safety shortfalls of seatbelts, are following the same basic trajectory, except faster. Cars and trucks are generally safer now, but also more expensive, which raises the cost to all consumers. The actual consumer may or may not have chosen to adopt these particular technologies, but is forced to pay the safety surcharge anyway. The cost of technologies often drops over time as production costs drop; however, this is not always the case. For example, advances in medical technologies have improved health care outcomes, but generally operate to increase the overall cost of health care dramatically. As the cost of health insurance rises and more people find themselves without insurance, the cost of health care continues to rise to offset treatment for those who cannot pay.

In some cases, it is not even the consumer who benefits. Genetically modified foods (GMFs) have obvious potential benefit to those holding the patents. Farmers stand to benefit from the promise of disease, insect, and drought resistance, higher yields, and other enhancements. However, the consumers may or may not benefit from GMFs at all, and many are worried that the costs of health and environmental uncertainties outweigh any benefit that might accrue (Mehta and Hunt 2006).[13]

With respect to science, new discoveries or theories can prove disruptive to deeply held beliefs and cultural foundations. One need not look further than the debates raging around human evolution and intelligent design or creationism to see how socially divisive science can be when it is introduced into the public sphere without a strategy for meaningful dialogue on the deeper implications and meanings. A narrow focus on technological novelty, competitive pressures for new bells and whistles, and the attitude that we should create something just because we can, all show a lack of regard for impacts in the social spheres. Again, given the scope of potential change promised by nanoscience and technology proponents, any approach to ethical considerations must be acutely aware that the societal changes will manifest well beyond the immediate insight of the science or functionality of the technological application.

2.4 Scientists and Engineers as Moral Agents

The discoverer of an art is not the best judge of the good or harm which will accrue to those who practice it.

—Plato (Phaedrus)

All individuals are capable of moral and immoral acts. Moral agency is generally taken to refer to individuals who are self-aware and self-interested. A moral agent is capable of some level of rational thought, including the ability to weigh choices and form self-interested judgments. As suggested by Plato above, moral agents may not always make the best choices for any number of reasons. They may not have all the necessary information or time they need to properly weigh their options. They may lack the ability or will to consider the situation from more than one viewpoint. Narrow self-interest, conflicts of interest, or other coercive factors may push them to disregard other moral commitments. In the case of "discoverers" in science and technology, personal investment and belief in their own creation may cloud their objectivity or willingness to question their motives or the potential harms of their creation. Yet, we do expect someone who is capable of making a moral judgment to make the effort to do so.

This expectation is particularly high for those members of society who are associated with a professional discipline. Professions such as medicine, law, engineering, and science are privileged on the basis of the expert power embedded in their training and the legitimate power they are afforded in order to practice their profession in society. Their ability to practice their profession relies, in part, on their willingness to faithfully adhere to certain attitudes and standards of behavior. For example, when one is admitted to a hospital for medical care, you can reasonably expect that the physicians, nurses, and other clinical providers who care for you are properly educated and licensed. You want them to be current with the best evidence-based practice in their field. You also expect them to provide care according to their designated scope of practice, to decline to participate in care for which they are not properly trained or competent, and to follow all appropriate regulations, policies, and procedures intended to assure your safety and well-being while under their care. Finally, you can expect to be told the truth about your condition and informed of the risks of your care. For the practicing health care professional, these expectations become ethical obligations that place you, as the patient, at the center of their practical and moral duties.

For engineers, a similar role fidelity exists insofar as they are charged with the design and testing of objects that meet a specific human need with effectiveness and efficiency while also attending to the safety and well-being of those users (Emison 2004).[14] Rather than a single patient as the focus of their immediate duty, engineers must meet the needs of their client and whatever users are likely to adopt a particular product of their design. However, they have similar duties regarding proper training, competency, currency in the field, observance of applicable regulatory requirements, and quality and safety measures (Uff n.d.).[15] They also have a duty to inform users of risks.

Not all scientists serve the interests of individuals in society quite so directly, as noted earlier in the discussion of the perceived difference between basic or pure science and applied science. They do not have an obvious professional-client relationship and, for this reason, even their status as a profession is debated (Chadwick 2005).[4] D'Anjou (2004)[16] suggests that "design-grounded" professional disciplines, such as engineering and medicine, are engaged in action that effects direct change in the human condition, thus forcing them to contemplate the moral dimension of their impact on society. Science, on the other hand, is primarily concerned with explaining what already exists with a certain lack of engagement needed to maintain the proper objectivity. Perhaps for this reason, science ethics tends to focus primarily on issues that are important, but of secondary interest to other professions. The ethical lens of science focuses on issues that undermine scientific credibility, such as fabrication or falsification of data, failure to report negative findings, giving appropriate credit to authors, and conflicts of interest in the funding sphere (Bird 2002).[17] However, it can certainly be argued that the line between basic and applied science is not always clear, certainly in the moral dimension. For example, current research efforts in nanoscience are clearly driven by the potential for application. As the relationships between science, engineering, and commercial enterprise continue to merge, it becomes harder to separate the work of the scientist from the work of the engineer and the entrepreneur. This blurring of lines is part of what has contributed to a general call for greater public accountability of scientists and increased public engagement earlier in the research and development process.[4]

Individuals all wear multiple hats when it comes to ethical obligations. In the language of ethics, this is termed conflicting loyalties or conflicts of fidelity. In any particular situation, you may find yourself needing to be loyal to your employer, your colleagues, your profession, your family, your community, and yourself. These multiple loyalties often challenge us to consider what we do, and the manner in which we do it, in light of each different stakeholder.[14] For example, the immediate interests of the company for which you work may conflict with what you believe to be legitimate concerns regarding public safety. In another example, public funding priorities may violate your sense of how resources should be allocated to serve the greater good. On the flip side, your personal financial well-being could be at stake if funding priorities shift from one direction to another.

We don't have good systems and processes for dealing with such conflicts in modern organizations and social institutions. However, the discomfort felt when such conflicts arise is evidence that scientists and engineers do have a role and a responsibility to take the societal implications of their work seriously. By using the language and concepts of ethics in a productive and collaborative dialogue, the evolving opportunities of nanotechnology can be realized while also minimizing the inevitable disruptions and harms of technological advancements.

COMMENTARY

The Public Engagement as Scientific Responsibility

Daniel Sarewitz and David H. Guston

Novel, complex, uncertain, and public, nano-scale science and engineering (NSE), or nanotechnology, may be the most post-normal (Funtowitz and Ravitz 1992) technoscience yet. This assessment comes not merely from the community of scholars in science and technology studies, which was in many ways late to recognize NSE's interest (Bennett and Sarewitz 2006), but from an increasing number of governments and private sector organizations calling for and organizing activities aimed at informing the public about NSE, and eliciting public views about it. In the United States, the 21st Century Nanotechnology Research and Development Act of 2003 (Public Law 108-153) authorized the integration of "public input and outreach...through mechanisms such as citizens' panels, consensus conferences, and educational events" into nanotechnology R&D (Fisher and Mahajan 2006). The U.S. National Science Foundation (NSF) has pursued this charge by funding a Nanoscale Informal Science Education Network (NISE Net) and two centers of nanotechnology in society, for a total of more than $30 million over five years. In May 2006, the National Nanotechnology Coordinating Office hosted a workshop on "Public Participation in Nanotechnology" that drew 120 participants and turned many away. Yet, at the core of even this most progressive of U.S. science and technology policies lays an uncertainty about precisely what the rationale for and importance of public engagement is.

Scholarship on public engagement in science and technology decision-making has most often formulated its necessity as a normative claim on behalf of the public. Occasionally, engagement is justified as a pragmatic claim to include local knowledge for the co-production of the best available information for decision-making. In nanotechnology, public engagement is increasingly rationalized as a way to soothe the slumbering beast of the public, lest it be roused and derail new industrial technologies, as some interpret the GM food experience, or nuclear power. Less often has public engagement been conceived of as a scientific responsibility. When it has, that conception has tended to err on the side of the deficit model of public understanding of science—which holds that knowledge is a fundamental prerequisite for participation—rather than an engagement model that sees rights or interests as a sufficient stake.

But the political economy of the scientific enterprise renders concepts of individual responsibility and ethical conduct, while quaintly necessary, largely moot at the societal level. Researchers themselves do not feel as if they can be effective moral agents (Berne 2006), and the individual scientist has little, if any, meaning in the global innovation system. There are millions of such scientists and more entering labs each day. The scale, complexity, aims, and outcomes of the enterprise may be completely opaque to individual scientists, but like soldiers on a chaotic battlefield, this fog does not compromise their ability to perform their individual tasks. To locate scientific responsibility primarily in the individual scientist is thus to render it meaningless. It's not that we shouldn't expect scientists to behave ethically, or to "communicate" with the "public." Rather, it is just such behavior, even when considered cumulatively across a community of scientists, that has little or no connection to the dynamics of innovation and its impacts on society. To put it more starkly: We can populate our innovation system with scientists who pursue truth, behave ethically, and communicate effectively, and still end up with outcomes that we cannot manage, or that challenge our values and principles. Is there then some notion of scientific responsibility that can play a useful role in confronting the waves of technological transformation that seem likely to inundate society in the next few decades?

The last several decades of social science scholarship on science and innovation provide powerful insights for confronting this question. We know, for example, that scientists negotiate not only with nature to advance knowledge, but with each other, with their funders, with politicians, corporate executives, and with various publics. We know that the directions and velocities of science reflect decisions made by people, and decisions emerge within a social context. We know that context is strongly embodied by the institutions where science is conducted and planned. These understandings have not yet been brought to bear on new notions of scientific responsibility, but if we are to have any prospect of guiding the accelerating technological change that engulfs us, now is the time to consider how we might accomplish this task.

Where might we intervene if we are to move toward a meaningful notion of scientific responsibility—and thus, accountability? Responsibility must be located in the processes by which decisions about science are made and implemented in the institutions of public science, rather than in the motives and norms of individuals who conduct science. Responsibility in science must manifest in a capacity for reflexiveness—for social learning that expands the realm of choice available within public research institutions. Above all, scientific institutions, broadly defined, will have to build the capability for scientists and technologists, as well as research planners and administrators, to understand where their research agendas come from, what interests are embedded in those agendas, and who supports them and opposes them, and why. Institutions need not only to enable conscientiousness (Berne 2006) on the part of scientists about the contexts in which they do their work and make their decisions, but they need to empower that conscientiousness with persistent collaboration with social science and humanities researchers, open and informed engagement with various interest groups and publics, and unflinching reflection and deliberation about motives and values. These capacities need to be built into research institutions. In the ideal, the support and expansion of conscientiousness can help signal emerging problems, support anticipatory governance, and enable better choices to be made about the directions and pace of knowledge creation. This creative process emerges not from any one scientist behaving responsibly, but from all of them doing so through their institutions.

References

Bennett, I. and D. Sarewitz. 2006. "Too Little, Too Late? Research Policies on the Societal Implications of Nanotechnology in the United States." *Science as Culture* 15(4):309-325.

Berne, R. W. 2006. *Nanotalk: Conversations with Scientists and Engineers about Ethics, Meaning, and Belief in the Development of Nanotechnology.* Mahway, NJ: Lawrence Erlbaum Associates.

Fisher, E. and R. L. Mahajan. 2006. "Nanotechnology Legislation: Contradictory Intent? U.S. Federal Legislation on Integrating Societal Concerns into Nanotechnology Research and Development." *Science and Public Policy* 33(1):5-16.

Funtowicz, S. O. and J. R. Ravitz. 1992. "Three Types of Risk Assessment and the Emergence of Post-Normal Science." Pp. 251-73 in S. Krimsky and D. Golding, eds. *Social Theories of Risk.* Westport, CT: Praeger.

Daniel Sarewitz is the Director of the Consortium for Science, Policy & Outcomes, and Professor of Science and Society with Arizona State University's School of Life Sciences and School of Sustainability. His work focuses on understanding the connections between scientific research and social benefit, and on developing methods and policies to strengthen such connections. His books include *Shaping Science and Technology Policy* (UW Press 2006); *Living with the Genie: Essays on Technology and the Quest for Human Mastery* (Island Press 2003); *Prediction: Science, Decision-Making, and the Future of Nature* (Island Press 2003); and *Frontiers of Illusion: Science, Technology, and the Politics of Progress* (Temple University Press 1996).

David Guston is the Associate Director, Consortium for Science, Policy & Outcomes and Professor of Political Science at Arizona State University, as well as the Director of the Center for Nanotechnology in Society at Arizona State University. He is the author of *Between Politics and Science: Assuring the Integrity and Productivity of Research* (Cambridge University Press 2000), winner of the American Political Science Association Don K. Price Prize. His other books include *Shaping Science and Technology Policy* (UW Press 2006); *Informed Legislatures: Coping with Science in a Democracy* (University Press of America 1996); and *The Fragile Contract: University Science and the Federal Government* (MIT Press 1994).

2.5   The Business Community and Corporations as Moral Agents

Under the law, corporations are chartered as citizens of the state and, as such, hold all of the rights and responsibilities of an individual person. As an entity, the corporation can then be held accountable for its business activities, including decisions, actions, and outcomes along with the conduct of its employees. In other words, a corporation is recognized as possessing agency. The corporation is a special case of a moral agent in that it is responsible to society for its decisions and practices.

Obviously, the corporation itself does not make decisions or take action apart from the individuals within the organization who have the power and ability to do so. Therefore, it is particularly important that those individuals be willing and capable of building an organization that can recognize, deliberate and act upon moral issues as they arise in the normal course of business. What is called for might be characterized as an organizational conscience and it stems from a variety of sources.

The first source of conscience includes the values operant within the organization. An organization embodies three distinct categories of values. The first category is represented by the cumulative personal values carried into the organization by its leaders and employees. Although derived external to the organization, individuals do not simply check their values at the door each morning. Rather, their personal values inform how they interact and interpret activities and intentions within the organization. The second category might be termed the espoused values of the organization. These are the values found in formal values statements, employee orientations, and presented as the public face of the organization. Finally, the third category includes the actual lived values of an organization. Lived values are apparent in how the organization actually operates. For example, while human resource policies may embody one set of espoused values, the actual human resource practices may be quite different (lived values). How resources are allocated, how suppliers are treated, and the extent of financial transparency are all measures of the lived values of an organization.

The second source of conscience is focused within the leadership team. As the primary holders of power and influence within the organization, the motives, intentions, and goals of the organization's leadership drive everything from strategy to operations. Leadership embodies much of the will of the organization when it comes to moral agency. Leaders who are overtly self-interested may be less inclined to confront ethical issues that may reflect badly on them or undercut their personal financial interests. Leaders who lack the courage to confront issues are equally unsuitable as the driving force of an organizational conscience. More commonly, leaders simply lack the moral imagination and confidence to recognize and confront morally questionable situations as they occur.

This need for an organizational will leads to a third source of organizational conscience in the form of culture. The intersection of personal, espoused, and lived values shapes the foundation of an organization's culture and determines how people within the organization are likely to respond when confronted with a moral choice. In an organizational culture where debate over ethical practice is encouraged and nurtured, issues will rise to the surface and are more likely to be resolved effectively (Johnson 2007).[18] In an organizational culture where conflict is shunned and debate shut down or punished as disloyal, then ethical questions remain unasked and unanswered. Unethical practices are likely to be implicitly endorsed and the culture begins to actually cultivate unethical behavior.

You can readily recognize this dynamic in the large corporate scandals, such as Enron, where many people knew that what they were doing was either overtly illegal, marginally legal, and/or irresponsible with respect to their fiduciary duties. Yet, so long as the profits appeared to be rolling in, employees and investors essentially looked the other way as a matter of the corporate culture. In the case of Enron, regulators, economists, and policy makers largely looked the other way as well.

This raises the final source of organizational conscience in the form of structure and process. Organizations with a strong moral conscience have structures and processes in place that force moral reflection on the business and its practices.[18] Questions regarding the impact of the company's business practices on stakeholders, including employees, suppliers, distributors, customers, and the local community are raised as a matter of routine. Environmental impacts are also routinely considered along with short- and long-term implications of new products and processes. Policies and procedures intentionally reflect the espoused values of the organization and are enforced. Avenues exist for employees to express concern with corporate practices; managers feel free to dissent when decisions seem to violate moral tenets; and, whistle-blowing is endorsed with full employee protection.

Over the last couple of decades, there has been increasing attention paid to corporate ethics and responsibility as a matter of public accountability and good corporate citizenship.[18] The relationship between business and the public trust has been damaged by the prevalence of high profile, actual and perceived cases of corporate misconduct; political scandals that implicate corporate interests; the negative impacts of downsizing and outsourcing on the workforce; and, opposition by business to everything from raising the minimum wage, to environmental standards. Business ignores this reality at its peril when it comes to potentially controversial technologies. Class action lawsuits in the pharmaceutical industry are likely as much a symptom of public dismay at the lack of transparency and the single-minded focus on corporate profits as they are any actual wrongdoing.

For nanotechnology and the related technologies it may enable, the implications are becoming increasingly clear. Mounting attention is being paid to public perceptions of science in general and nanotechnology in particular. While public attitudes have generally remained favorable toward nanotechnology

(Bainbridge 2002; Cobb and Macoubrie 2004; Siegrist, Keller, Kastenholz, Frey, and Wiek 2007),[19,20,21] the same cannot be said for public attitudes toward business. In a widely cited study of the public perceptions of nanotechnology by Cobb and Macoubrie (2004),[20] more than 60% of respondents stated that they did not trust business leaders to have the ability or willingness to minimize any risks of nanotechnology to humans. In fact, only 5% admitted to having a lot of trust. Furthermore, the greater the perception of risk, the lower the level of trust. Cobb (2005) found that when more specific information on nanotechnology risks is provided, trust in industry leaders drops even lower. Interestingly, and importantly, the opposite is not true. Information about the benefits of nanotechnology does not work to increase trust.[22]

While such studies obviously provide no evidence of improper intent or widespread wrongdoing on the part of corporate and industry leaders, they do demonstrate a lack of trust that broadly affects any effort to introduce new technologies to the marketplace. Further, this lack of trust may well increase the likelihood that the public will respond negatively if unexpected risks and harms do surface at some point.

Calls for voluntary efforts on the part of industry are one means of rebuilding a foundation of trust. For example, Balbus, Denison, Florini, and Walsh (2006)[23] suggest that, in the absence of adequate regulatory provisions for nanomaterials that are ready or nearly ready to enter the market, industry and individual corporations should assume the leadership role in establishing and implementing "corporate standards of care" for managing the potential risks of a nanomaterial across the product's life cycle. To do so, they further advocate engaging an array of stakeholders in this effort, including labor groups, consumer advocates, community groups, and health and environmental organizations.[23]

2.6   Policy Makers and Regulators as Moral Agents

With the issue of trust, we also confront the role of policy makers and regulators as moral agents. Public policy and administrative law are intended to both represent and safeguard the public interest. Regulators are charged with assessing and then acting on behalf of the best interests of society at large. To do so, they must strike a difficult balance between what is allowed and what is limited. They fail this commitment in any number of ways. Over-regulation, under-regulation, lack of enforcement, overzealous enforcement, ideological or political agendas that favor one party over another, and simple lack of technical understanding all represent practical and moral failures of policy makers and regulators to serve and protect the public interest.

Much as we saw with public attitudes toward business, studies are not optimistic about the role of government in the assessment and regulation of new technologies. Siegrist and his colleagues (2007) studied the difference in attitudes between laypeople and experts in Switzerland. For laypersons, they found that higher levels of trust in governmental agencies that are responsible for regulating nanotechnology reduced the perceived risks, especially when combined with increased familiarity with potential benefits. However, experts, whose training and experience make them more knowledgeable about the hazards of nanotechnology than laypersons, were generally found not to be influenced by perceptions of benefit or general attitudes toward technology. Instead, their risk assessments were strongly influenced by confidence or lack of confidence in government agencies. Experts who expressed confidence in government agencies to protect against health hazards perceived lower levels of risk than those experts who were not confident in government bodies. In particular, these researchers warn against the possible effects of "a preventable event with significant negative consequences." They contend that perceptions of a lack of concern for public welfare on the part of the government or industry are likely to result in decreased acceptance of nanotechnology generally. Their recommendation is to promote voluntary initiatives and regulations that prevent untoward consequences while also enhancing trust in governmental agencies.[21]

Macoubrie (2006) also found low levels of trust in government, and extending to industry, to manage the risks of nanotechnology, with higher respondent educational levels being correlated with lower levels of trust.[24] Participants wanted more testing of products and better access to information by the public. They did not generally feel voluntary industry standards to be sufficient. This generalized mistrust is especially interesting in light of her findings that most respondents anticipated major benefits from nanotechnology and generally did not support a ban on nanotechnology products.

Commenting on the apparent decline in trust in both science and the ability of governments to adequately regulate it, Mehta and Hunt observe:

> Trust is difficult to build and easy to lose. The development of innovations in biotech has been hampered by this lack of trust due to public concerns about the adequacy of the regulatory process, its openness and transparency, and potential conflicts of interest arising from government, industry, and university partnerships. Low levels of trust make technologies inherently unstable from a social perspective, and often lead to overreaction.[13]

Bolstering this basic pessimism are observations that current policy and legislative efforts are decidedly mixed in their message. Fisher and Mahajan (2006)[25] analyzed the motivations and strategies implicit in the eleven Program Activities for the National Nanotechnology Program as authorized by the 21st Century Nanotechnology Research and Development Act (2003). They grouped the activities into the categories of techno-scientific, global-economic, and ethical-societal and discovered there to be two somewhat competing and potentially opposing ends. The first is rapidly accelerated development and deployment of nanotechnology, while the second is responsible development. These authors contend that the means for balancing rapid deployment with adequate attention to societal concerns is not clear. They conclude:

> The congressional response to the perception of opposing pressures surrounding nanotechnology has resulted in antithetical legislative language which, in turn, prescribes mixed models of R&D. Depending on how it is implemented, the Act could emerge as a shrewd piece of legislative rhetoric, reducing societal research and related activities to a sideshow in order to push rapid nanotechnology development past a potentially wary public, or as a tool for ushering in a prudent new paradigm in technology development; this would require a radical reevaluation of the relationships among technological activities, their products, and the associated outcomes.[25]

Others have also acknowledged a relative lack of action on the part of federal agencies to anticipate regulatory and funding needs. For example, Balbus, et al. (2006) call for urgent action to increase risk research, citing the call by experts at a 2004 NNI workshop to increase the $10 million dollar level of risk related research funding to $100 million. They believe regulatory programs in the United States are particularly susceptible to having nanomaterials "fall between the regulatory cracks" due to lack of regulation or under-regulation.[23] This lack of oversight increases the likelihood that specific risks may go undetected until adverse effects are widespread.

The combination of public mistrust, mixed messages from government, and a perceived lack of meaningful action do not bode well for government's role in the ethical dialogue regarding nanoscience and technology, particularly here in the United States. Recent revelations regarding the possibility that government science reports were vetted by former industry lobbyists joins a long list of complaints against an Administration that has consistently resisted transparency and disclosure on a range of issues. Better modes of disclosure and a more engaged, two-way communication with stakeholders at many levels of government will be necessary in the future if productive consideration of the ethical questions surrounding these powerful new technologies is to occur.

2.7 Challenges of Pace, Complexity, and Uncertainty

Emison emphasizes the ethical challenges confronted by engineers in the face of growing complexity. He attributes this added complexity to several factors, including multiple claimants with conflicting interests, novel settings, expanded demands, and the need for new skill sets beyond technical expertise.[14] Engineers increasingly find themselves answerable to a wider range of stakeholders than just their clients, including regulatory agencies, public interest groups, and even the news media. In response to the general shift in the marketplace away from traditionally structured organizations and relationships, they find themselves practicing in novel situations that require different economic, political, and interpersonal skills to navigate. The demands related to their design objectives have expanded to include everything from environmental issues to societal considerations, such as compliance with the Americans with Disability Act.[14]

This same observation holds true for scientists, business leaders, and policy makers. Scientists are increasingly tied directly to the product research and development process in ways that force them to be more accountable for the practical outcomes of their research. Competition for funding and other resources, increased media scrutiny, and pressure to produce in a fast-paced competitive marketplace are all factors that complicate the moral dimension of the scientist's relationship to the larger social order. Likewise, businesses are also forced to respond more and more quickly to a wider range of stakeholders, and in a changing marketplace awash in uncertainty and unpredictability. Finally, regulators and policy makers are generating much of the complexity in their efforts to respond to environmental, health and safety, economic, legal, political, and other concerns in a fast-paced, rapidly changing, and highly uncertain global environment.

The upshot of this dynamic and highly reactive environment is that it becomes easier and easier to push ethics off as a longer term ideal rather than an immediate practical demand. Nowhere may this prove truer than in a situation in which advances in nanoscience and technology initiate major and rapid change across multiple industry sectors simultaneously. However, failing to address moral concerns related to risk and other forms of societal disruption does not make them go away; nor do such issues tend to solve themselves. In fact, these issues are far more likely to become more and more divisive or demanding over time, often resulting in severe economic consequences or a crisis in public trust. Asbestos is a common example of how costly failing to anticipate risk or moving rapidly to address it can be. A more effective approach is to recognize ethical analysis as a practical investment that pays off by helping to minimize the disorienting effects of pace, complexity, and uncertainty.

Emison goes on to recommend the philosophical view of pragmatism as the proper foundation for engineering ethics in the current environment.[14] In fact, certain assumptions of pragmatism are well suited to the demands of ethical analysis and decision-making under conditions of rapid pace, increasing complexity, and high uncertainty. One core assumption of pragmatism is that choices are highly contingent on particular circumstances. In the face of complexity, it becomes harder and harder to apply a limited set of "ethical" responses to emerging and often novel circumstances. A "correct" response in one circumstance may prove to be wrong or ineffective in another similar circumstance. Furthermore, pragmatism assumes that decisions in one social setting may provoke very different meanings in another. While this constitutes a challenge and potential conflict on one level, on another it means that divergent views can be used to expand, negotiate, and refine more sophisticated or finely nuanced solutions and ways of interpreting moral obligations under conditions of complexity. A final assumption of pragmatism is somewhat more problematic in that it contends there are no universal principles or guidelines for human action. In fact, the ability to find common ground is the core skill in conflict resolution and, at the very least, a set of broadly applicable, guiding principles can help frame a conflict or analysis in a way that helps to prevent the situation from devolving into narrow self-interest, defensive power struggles, or the squeaky wheel syndrome. At the same time, the pragmatist's emphasis on the actual experience and particular context of a situation provides a foothold that prevents dialogue from becoming too abstract and "academic" to be of practical use.

2.8   A Matter of Trust

In this country, the formal calls for public dialogue regarding nanotechnology are, in large part, a response to the general fear that public suspicion and backlash may waylay U.S. economic development, global technological supremacy, and even national security itself (Fisher and Mahajan 2006; Berube 2006).[25,26] However, implicit in such calls is also a recognition that a necessary partnership between the political, economic, and public spheres is absent or not functioning as well as it should. Naturally, such dialogue is notoriously difficult to facilitate because the parties tend to represent different sets of interests and priorities. In other words, they tend to speak somewhat different languages that are a bit suspect within the other camps. A workforce concerned with job security and a living wage has a decidedly different focus than investors and corporations under pressure to meet quarterly profits, or regulators charged with overseeing all facets of the economy at a global scale, as well as consumer safety, environmental protections, and other public concerns. Although the practical concerns are many, the inherent conflict is essentially one of values, and our perceptions of which outcomes are most valuable to us as individuals and the community as a whole. Therefore, public dialogue regarding any aspect of nanotechnology is, at its core, an analysis of ethical priorities and responses.

Siegrist and his colleagues (2007) provide an interesting insight into the nature of trust between the public, science, and the institutions of business and government. In reflecting upon the conclusion of their study, they point to past research that suggests a difference between trust and confidence. They define confidence as largely "based on experience or evidence that certain future events will occur as expected," while trust is based on "shared values." They believe that, in their research, the more technically and experientially informed experts were expressing confidence or a lack of confidence in their views on whether the government would adequately address concerns. Laypeople, with less knowledge and experience, were expressing either trust or distrust that the government shared their values.[21]

The concept of shared values, although prevalent in the business and organizational literature, is not a particularly well-defined concept. In discussing the sources of corporate values earlier, I alluded to at least three: personal values, espoused values of the organization and its members, and the actual lived values of the organization and its members. One could assert that shared values are most likely to reside at the conjunction of these three sources. A similar assertion might then call for an examination of how best to align the values of science, business, governance, and the public in an effort to build the foundation of shared values upon which a working level of trust can be built.

In a thoughtful commentary from late 2005, communications researcher Susanna Priest[27] reflects at length on the nature of trust as "an expectation for the behavior of others." She contends that trust should not be minimized as a commodity of which one needs just enough, noting that people who lack trust in scientists, engineers, and other leaders may simply have learned not to trust them over time. She points to a laundry list of technological missteps, including Bhopal, Chernobyl, Three Mile Island, Dalkon Shield contraceptive implants, thalidomide, DDT, Agent Orange, the U.S. space shuttle program, the sinking of the Titanic, the Exxon-Valdez oil spill, and the Tacoma Narrows bridge collapse. She warns against ignoring these large-scale and other smaller failures, repeating the same mistakes and undermining public confidence in the process.

> Suspicion (that is, lack of trust, at least under some circumstances, for some people, regarding some technologies) is certainly a reasonable public response to this history, even though the level of suspicion that is justified in a given case might be debatable—that is why we need the debates. The phrase "healthy skepticism" is not an oxymoron. We need to reassure people that the health and environmental implications of nanotechnology and other new technologies are being carefully considered, as are the ethical dimensions, *and we also have to actually consider them*, not dismiss these concerns as ("merely") emotional.[27]

This very practical description of the nature of trust can also be interpreted as a call to action by the scientific, business, and government communities to authentically engage and reflect on their own intentions, goals, obligations, and approach to serving the public trust. On the other hand, the public and its representatives also owe a certain obligation to be informed and reasoned in its assessments and conclusions, something to which we will return in later chapters.

Following a brief presentation I did a couple of years ago at a conference aimed at engaging the business community in the societal implications dialogue, I was confronted by a member of the audience who accused me of "interfering with his livelihood." My presentation was a rather innocuous call for a balanced dialogue between the obligations of the scientist and society's obligations to the scientist. Yet, the fact that I used the word "ethics" was an affront to this gentleman who felt I was imposing my values on his ability to do business. After calming him down somewhat, I realized that I had simply been the straw that broke the camel's back near the end of a long day in which he had selectively heard only what seemed to him to be anti-business views, while ignoring the myriad of supportive statements. Instead of a call for dialogue, he heard only personal attacks. Instead of an offer of partnership, he saw only additional barriers to the already considerable challenges of successfully bringing new products to a highly competitive global marketplace. In fact, he clearly, and rightfully so, viewed his business activities as positive contributions to society. Unfortunately, he also perceived them as under-appreciated and under open attack by people who simply do not understand business.

This encounter was important to me. Although as an ethics instructor I am disciplined to be sensitive to multiple viewpoints, I was startled with the ferocity of his self-defense and have since observed such defensiveness in many encounters with business and governmental interests. Believe me, the fastest way to cramp the discussion at a lunch table is to introduce yourself as an ethicist! Therefore, I want to leave this chapter with a strong statement of support as well as caution.

Science, engineering, business, and government all represent vital interests to the current and future well-being of society. It is the nature of society to change in response to nearly everything and a great deal of that change occurs in a positive, productive direction. Science, technology, and commerce will be the source of many positive societal impacts now and into the future. To operate from an assumption that societal impacts research is focused only on risks or negative potentials is to undermine its value to the entire scientific and technological enterprise. To assume that any discussion of ethics is intended to point out only what is wrong about technology, or with those who advocate for it, is to misconstrue the very nature of ethics, which seeks first and foremost to promote the good in some form.

At the same time, what is implied by the term societal and ethical dimensions is an enormous range of issues, much of which lacks explicit definition. Furthermore, meaningful analysis will cross many disciplines and stakeholders. Yet, there is no real infrastructure in which such interdisciplinary collaboration can easily emerge and even the idea that natural scientists and engineers should be working collaboratively with the social sciences and humanities to proactively assess societal impacts and ethical implications is very new and not entirely well accepted. The need for infrastructure and acceptance implies a fundamental change in the culture of science and technology, as well as its relationship to both policy making and public dialogue. All that is called for in this chapter is an enhanced understanding and appreciation of longstanding assumptions, and the complexity inherent in the relationships that exist between stakeholders when considering ethics and science.

2.9 Questions for Thought

1. There is no generally accepted code of ethics for scientists. If there were, what principles or guidelines might it require?

2. One of the outcomes of the NNI has been an unprecedented collaboration between university-based researchers, business interests, and the government. What are the potential ethical issues and conflicts inherent in such a collaboration and how might they be addressed?

3. The ability to manipulate matter with precision at the atomic level will afford human beings an unprecedented power to control their world. This power may differentiate nanotechnology from prior technologies, generating controversies between science and religion similar to those we currently see related to human evolution, creationism, and intelligent design. What strategies might help facilitate a more productive dialogue and response?

3

Societal Impacts and Perspectives

> An idealist believes the short run doesn't count. A cynic believes the long run doesn't matter. A realist believes that what is done or left undone in the short run determines the long run.
>
> —Sydney J. Harris

3.1 Fundamental Transformations

When applied to technology in general, Harris's comment makes a good deal of sense. Decisions made, or not made, and actions taken, or not taken in the present can follow us long into the future. If we fail to make a decision or to act when confronted with the opportunity, we are every bit as ethically accountable as we are when we do make a decision or take action. A bit too much optimism can lead one to justify unreasonable risks in the immediate or short terms. At times, such optimism can make it difficult to even recognize the potential pitfalls of an endeavor. The cynical view, on the other hand, can leave one without the energy or motivation to move ahead responsibly, even when the potential benefits are great, or the potential problems obvious. In either case, the success of a project is jeopardized by incomplete planning and the inability to respond effectively to both opportunity and risk.

In their 2001 report, Societal Implications of Nanoscience and Nanotechnology, Roco and Bainbridge[1] would seem to be calling for the attention and action needed in the short-term in order to ensure success in the longer term.

> Over the next 10-20 years, nanotechnology will fundamentally transform science, technology, and society. However, to take full advantage of opportunities, the entire scientific and technology community must set broad goals; creatively envision the possibilities for meeting societal needs; and involve all participants, including the general public, in exploiting them.[1]

However, six years later, we have not even managed to fully define the terminology needed to explore the issues. Unfortunately, this can lend legitimacy to any questions that have been raised about the motivations and intentions of parties on all sides of the debate.

We've already considered the lack of a consistent definition of nanotechnology itself and the problems with framing it as either evolutionary or revolutionary. Naturally, the lack of clear definition and agreement on key technical terms can hamper everything from reporting and interpreting research, to generating standards, to regulating safety, to informed investing, to appropriate consumer marketing. Likewise, the lack of clarity in descriptive language can lead to misrepresentation, misperception, miscommunication, and misdirection. Imprecise description and categorization complicates meaningful investigation and dialogue, particularly in the realm of societal impacts.

Take, for example, the phrase *fundamental transformation*. What does it mean? How does one define a fundamental transformation? Together, the terms suggest a significant change in the very essence or core structure of something; however, society is in a state of relatively constant change, so which changes will meet the threshold of significance to qualify them as fundamental transformations? Will we just know it when we see it? When will it start? Will we even be aware it is happening? Is there a tipping point at which "normal" innovation and change suddenly become fundamental and transformative? How will we recognize that point? Does such a distinction obligate us to a higher order attention, action, or caution?

Notice in the earlier excerpt from Roco and Bainbridge how the tone of the language is basically positive and appeals to our essential optimism regarding innovation and progress. On the whole, their suggestion is that fundamental transformation is a useful thing. We are left with the impression that the worst that can happen is that we will fail to take advantage of, or "exploit" certain inevitable opportunities to meet societal needs. Implicit in the language is the essential goodness or desirability of those opportunities. In fact, the importance of involving all parties is to ensure everyone is on board and not impeding our forward momentum. There is a sense of predestined certitude and inevitability to their comments. It is not a matter of whether this happens, only when and how quickly.

Economist Jeremy Rifkin (1999),[2] referring specifically to advances in biotechnology, uses similar terminology but with a different flavor.

> Our way of life is likely to be more fundamentally transformed in the next several decades than in the previous one thousand years…. In little more than a generation, our definition of life and the meaning of existence is likely to be radically altered.

He goes on to critique the unqualified enthusiasm of scientists, corporate leaders, and politicians, suggesting that they either ignore historical experience or intentionally mislead the public with their optimistic predictions and assessments.

> Yet, if history has taught us anything, it is that every new technological revolution brings with it both benefits and costs. The more powerful the technology is at expropriating and controlling the forces of nature, the more exacting the price we will be forced to pay in terms of disruption and destruction wreaked on the ecosystems and social systems that sustain life.[2]

In this case, we sense a different, darker, and somewhat threatening tone. The emphasis is on the risks of the unknown and the lack of suitable caution when proceeding too quickly. Although acknowledging the potential for benefits, our attention is drawn to words like "disruption" and "destruction." Fundamental transformation is suspect and we find our instinctive resistance to change and fear of the unknown kicking into gear as we consider the possibilities of fundamental transformations that may not, in fact, enhance our lives and may even prove a danger to us or our children at some point.

In both cases, the essential bias of the writer's position suggests a different set of ethical considerations. In the case of Roco and Bainbridge, we might ask whether we might actually have a duty, that is, be ethically compelled, to actively support research and development of this technology in order to meet current and future societal needs, whatever those needs might be. Supporting the broadly targeted and rapid advance of technology, generally, and nanotechnology in particular, is a means of achieving the greater good. In fact, in an opening speech at a 2003 NNI workshop, Phillip Bond, former Undersecretary for Technology in the Department of Commerce, asserted this point directly by stating:

> Even if [nanotechnology] could be stopped, it would be unethical to stop it. In fact, I believe a halt, or even a slowdown, would be the most unethical of choices.[3]

He cites the following as moral imperatives: freedom from future pollution and the ability to repair existing environmental damage; feed the world's hungry; enable the blind to see and the deaf to hear; eradicate or protect against disease; and extend the length and quality of life through the repair and replacement of organs.[3]

However, in Rifkin's piece, the ethical challenge shifts to questions of avoiding outright harm, the obligation to balance benefits and burdens, consideration of the longer term consequences of our actions, and the potential for at least some technologies to threaten greater harm than benefit. In this scenario, our duty becomes one of caution along with the suggestion that we should question both our own motives and the motives of those in power.

In reality, a truly fundamental transformation is most likely to engender an array of both benefits and harms that are themselves open to multiple interpretations and complex interactions. The major flaw in these very short excerpts is that both seem to imply that we have a general agreement on how to define benefits and harms. Take, for instance, the ability to extend the human life span by simply replacing organs as they wear out or are damaged. Ask a random group of people if they think this is a good idea and some will readily agree that it is. For others, the "yuck" factor will come into play as they ponder whether this violates the natural order or divine intention. Yet, another group will immediately begin to calculate exactly how far their current retirement savings are likely to take them and how long they would have to take up space in the workforce in order to accommodate another 10 or 20 years of life.

We simply don't agree on what constitutes a benefit or harm, nor will it be easy to achieve such agreement since technological outcomes interact with society in complex ways. Some interactions reinforce each other, amplifying the positive or negative effects of a technological advance, while other interactions tend to offset and balance each other. When attempting to assess the relative value of any technology, and nanotechnology in particular, as a primary source of societal change, we are better served when the meaning of a term, such as *fundamental transformation*, is granted a certain broad objectivity. Manipulating the term within a preexisting bias tends to shut out the valid questions raised from multiple stakeholders and viewpoints. It oversimplifies and polarizes what should otherwise be both an objectively and subjectively balanced discussion.

In the remainder of this chapter, we will explore various efforts to define social and ethical implications of nanotechnology, along with a discussion of the intersection of science and human values in the broader public sphere. We will also examine the historical roots of societal impact activities, including the current attempt by the National Nanotechnology Initiative to address societal implications and the origins and role of the precautionary principle in current nanotechnology considerations. Finally, we will explore the role of citizen as moral agent.

3.2   Ethical and Societal Implications Defined

With the former points in mind, we can now turn to the question of what constitutes an ethical implication or a societal impact. There does not appear to be an operative definition of the combined terms ethical and societal implications or ethical and societal impacts. Any substantive advance in science or the introduction of a new technology is, in and of itself, a societal impact insofar as it has added to the body of knowledge or the technological capabilities of society in general. The depth and breadth of the impact may be small or large, but it is an impact nonetheless with moral and ethical implications. Impacts of any kind are rarely value neutral. The word *impact* itself means to strike, collide or otherwise effect. When something new gets our attention, we will interpret its effect on the basis of its perceived value. It will be seen to provide a benefit, harm, or some combination of benefits and harms. However, with respect to nanotechnology, we are again faced with a built-in bias in terms of how the language has come to be used. In this case, the bias tends to suggest that the technology itself is basically a good so long as we give adequate consideration to the ethical and societal implications, inferring that whatever falls into these categories of ethical and societal impacts constitutes a set of potential problems or negative concerns. If the implications were not problematic, there would be no particular need to consider them. Certainly, there would be no need to dedicate nearly 10% of the research funding to study them.

The lack of definition when using such sweeping terms, and the tendency to use the language with a certain built-in bias, adds to the overall fuzziness when attempting to discuss the meaning and value of nanotechnology. Lack of definition also makes it more difficult to set strategic objectives and priorities for continued research and development. There is no one framework or categorization that fully outlines what is meant by societal implications. There are a few categories that do consistently come up. For example, environmental concerns tend to be mentioned along with possible human health risks and issues related to privacy. Potential legal impacts are also frequently mentioned with emphasis on patent and intellectual property challenges. Yet, the sheer breadth of possible implications makes it difficult to structure any sort of comprehensive analysis. At the same time, if we do try to fit societal implications into discrete categories, we can fail to appreciate the complex interactions inherent in social systems.

There are at least four ways one might attempt to define and classify the broad category of societal and ethical implications. Schummer (2005)[4] insightfully describes societal and ethical implications within a framework of interest groups and the meanings generated by each group. A more familiar approach is to establish broad spheres of impact including legal, economic, and environmental. Another frequent approach is to classify impacts by common categories of concern, such as human enhancement, workforce disruptions, nanomaterial toxicity, and privacy. Finally, I would like to suggest a fourth approach in which ethics itself provides the framework for classifying societal implications.

3.2.1   Interest Groups and Meanings

Schummer's approach employs an intriguing collection of interest groups including science fiction writers; scientists; policy makers and science managers; business; transhumanists; the media and the public; and cultural and social scientists. Focusing on these groups in the United States, he argues persuasively that the lack of clear definition of nanotechnology itself leads to a debate that is more visionary than substantive. Science fiction writers, for example, often impart moral messages about the impact of technology in the social context. In many cases, the picture painted is dystopic, laced with themes of alienation and scientists gone amuck. The most obvious example, with respect to nanotechnology, is Michael Creighton's (2002)[5] take on the infamous grey goo scenario introduced by Drexler. However, there are also utopian visions of technology that focus on human enhancement in the form of expanded consciousness or otherwise simply tickle the reader's technological imagination with descriptions of radically new tools and materials. Notably, Schummer concludes that most science fiction narratives encourage moral reflection on a wide range of impacts, but without really taking a stand on the issues raised. They leave it to the reader to reflect, or not, on the deeper meanings of our emerging technological capacities. In fact, he and colleague Rosalyn Berne advocate using science fiction narratives to teach engineering students the societal and ethical implications of nanotechnology (Berne and Schummer 2005).[6]

Scientists, on the other hand, tend to focus societal implications narrowly on the specific, potential technological applications of their research and its industrial impact or, in the case of toxicologists and environmental scientists, narrowly on environmental and health risks. However, Schummer points to a group of mostly software engineers who dominate the popular bookshelves with radical visions of nanotechnology, in which dramatic social change is largely a positive result of revolutionary breakthroughs in science and engineering. Perhaps the most notable example is Ray Kurzweil (2005),[7] whose recent book, *The Singularity is Near*, proposes an almost unyieldingly optimistic view of the ability of nanotechnology, and its convergence with biotechnology and information technology, to help fully realize the next phase in human potential.[7]

Policy makers and science managers, who are accountable for funding decisions, naturally tend to draw on these more radical visions to project mostly positive economic impacts. The prospect of new products and expanded business opportunities allows government funding of research and development to be presented almost exclusively as economic development costs. Furthermore, such costs are particularly important if the U.S. is to retain its competitive edge in the global marketplace. A secondary, but related, theme is that of national security and our ability to retain military dominance as well. Schummer's analysis concludes that, for policy makers and science managers, societal implications are basically anything that needs to be addressed with the public in order to educate and prepare them for nanotechnology. These concerns include fears of self-replicating nanobots and related environmental concerns, along with an interesting theme of human enhancement, particularly as it is related to artificial intelligence and radically enhanced human cognition. Protection of privacy is another commonly cited concern in official government reports.

From the standpoint of business interest groups, nanotechnology represents a huge emerging market, generally quoted to be estimated at $1 trillion in the next few years and growing exponentially after that (Berube 2006).[8] Following the lead of policy makers and science managers, business interests promote the most optimistic views of nanotechnology as an economic engine, minimizing any other considerations in order to attract investors.

Interest groups that fall in or around transhumanism focus on the potential for nanotechnology to eliminate material need through radically new abilities to assemble materials and products on demand; cure or reverse all forms of human disease and aging; and redesign and improve the human body well beyond its current physical and mental capabilities. They argue there is a moral imperative to improve the species, pushing the technological envelope as far as it can go. In so arguing, there is a strong disregard for any level of risk and an acceptance of the disruptive or harmful effects of technology as simply being calculated risks that are worth the longer term benefits. In fact, many transhumanists will argue that if there are dangerous or inherently evil applications of nanotechnology, then we clearly have to make sure "we" get it first in order to ensure that humanity is protected from the apocalyptic possibilities raised by it falling into the wrong hands.

When discussing the media, Schummer points out that more than 75% of the articles in the general media deal with money or primarily the business side of nanotechnology. A few articles related to health and safety concerns have appeared in response to just a couple of specific studies. One example is a widely cited study by researchers in the United States that examined brain toxicity of buckyballs in fish. If one assumes that media coverage reflects public interests, then public concerns about societal implications are largely defined as economic, with some concern about health and safety. If, on the other hand, one assumes that the media, via selective coverage, also shapes what people know and think, then such public definitions of societal impacts are likely to be artificially limited.

The final interest group discussed by Schummer includes cultural and social scientists, whose primary interest is research and applying the varied expertise of their disciplines to the investigation of the complex relationship between science and technology, and the rest of society. One might conclude that this group is best capable of defining and classifying societal implications based on observations and theories regarding the interaction of technology and ethical norms, cultural values, belief systems, and perceptions of societal needs. Interestingly, Schummer points out that their research is primarily funded by, and in the context of, the intentions of policy makers, science managers, and other scientists. Therefore, there is an implicit pressure to more narrowly define societal implications and conduct research in a way that directs research findings into public education and the promotion of nanotechnology.

Schummer goes on to assess the relative impact each of these interest groups may have in the current or future debate regarding the ethical and societal implications of nanotechnology. First, he observes that the far more sophisticated meanings generated by cultural and social scientists presently have "no discernable impact" on any other group, and even more interestingly, neither do nanoscientists themselves. So much for firsthand expertise. Instead, Schummer proposes an alliance of "semantic leaders" that includes science fiction authors, visionary engineers, transhumanists, and business, whose mutually reinforcing meanings are likely to dominate emerging understanding of nanotechnology and its societal implications. On a final note, Schummer points out historical examples of new technologies in which "the visionary propaganda downplayed any possible problems or risks, denounced critical voices, caused fears and hostilities, and frustrated all those who were naïve enough to believe in the recurring visions."[4] In each case, there was an antiscientific backlash. He proposes that, with the combination of overly optimistic practical and economic projections and understated attention to the likely disruptions in various societal realms, nanotechnology is at similar risk.

I have spent what may seem like an inordinate amount of time summarizing Schummer's analysis because it so effectively illustrates the integral role of ethics in understanding even the basic definition of a societal implication. Each interest group is operating on a shared set of moral assumptions that must be acknowledged and understood as part of any larger debate or dialogue. The first assumption is that any particular interest group's view of nanotechnology has accepted some definition of a greater good at the foundation of their position, whether that be in the form of expanded scientific knowledge, economic prosperity, improved living conditions, stronger national security, or the next great leap in human evolution. Similarly, the intentions of each group reflect an underlying set of values. Admittedly, some individuals within each group may be narrowly driven by self-interest, while others are driven solely on the conviction of the inherent rightness or goodness of their actions. And, as is true with any broad categorization of persons, there are people within each interest group who hold widely divergent views. There are scientists who reflect deeply on the long-term social impact of their research; and there are business people and political leaders who have strong reservations about promoting technologies that are overly disruptive to the lives of their employees, or that challenge the deeply held beliefs of the voters who elected them. However, as a whole, these groups have each generally determined their primary goals to be of greater importance than competing goals, and this is essentially a values-based determination. Practically speaking, ethics as most of us know it, is predominantly a framework in which our core values are, or fail to be, lived out. Therefore, each interest group has a set of implicit and explicit moral assessments that underlie the meanings they associate with societal implications. Schummer's work points out the extent to which these moral assessments can conflict or be mutually reinforcing between interest groups, ultimately shaping both the practical and moral dimensions of the debate. In the end, ethical and societal implications are most likely to be defined as whatever the most powerful and influential coalition of interest groups decides they are, at least until potential untoward outcomes force a shift in the nature of that coalition.

3.2.2 Spheres of Impact and Categories of Concern

Somewhat more familiar approaches to defining societal implications either establish broad spheres of impact, including the law, economics, and the environment, or classify implications by common categories of concern, such as human enhancement, workforce disruptions, nanomaterial toxicity, autonomous nano-replicators, and privacy. Ironically, it is the attempt at specificity in these approaches that ends up being so misleading because it inadvertently minimizes the effects of pace, complexity, and uncertainty. Social spheres intersect and concerns are often highly interrelated and interdependent. By attempting to position implications in pre-defined categories or classifications, there is an implicit suggestion of boundaries or containment. Yet, we know technology is not a particularly well-contained commodity. Materials, methods, and products developed in one industry spill over into another, as do disruptions to the workforce, health and safety issues, influences on popular culture, etc. In fact, the pace of discovery accelerates this process, while the complexity of the interactions and the uncertainty of efforts at predicting those interactions render control via any mechanism increasingly difficult. Nonetheless, efforts to identify categories of societal impacts are at least a starting point for dialogue.

An example of the general effort to categorize societal implications by social sphere is found in the previously cited report of a workshop conducted in 2000 by the Subcommittee on Nanoscale Science, Engineering and Technology (NSET). The projected impacts were associated with the social spheres of education, technology itself, the economy, health care, the environment, the law, ethics, and a catch-all for "other" impacts (Roco and Bainbridge 2001).[1] Mills and Fleddermann (2005)[9] list the broad spheres of health and safety, medicine, privacy, economic, international, law, and education.

Within these broad spheres, one can then see specific categories of concern begin to emerge. However, a concern originating in one sphere can easily impact other spheres. In some cases, a troublesome impact in one sphere may be offset by a positive impact in another sphere. More commonly, problematic issues raised in one sphere generate problematic issues in others. Table 3.1 uses a set of commonly cited categories of concern and lists them based on spheres likely to be impacted. As you can see, very few issues stay solidly in one sphere, adding considerably to the challenge of assessing impacts in either practical or ethical terms. For example, a concern originating in the environmental sphere may have implications and potential impacts in the health and medical, economic, legal, and geopolitical spheres.

3.2.3 Moral Dimensions

A final approach might be to describe societal implications more broadly in terms of their moral dimensions and the various opportunities and threats they embody. Lewenstein (2005)[10] suggests this approach when he proposes a common frame holding together the disparate issues in nanotechnology. He contends that fairness, equity, justice and, in particular, power in social relationships serves as the ethical link between most issues that have been otherwise classified societal implications in the various spheres of impact discussed in the previous section. He suggests that the classification of issues

TABLE 3.1

Spheres of Impact and Categories of Concern

Environmental and Safety Sphere	Health and Medical Sphere	Economic Sphere	Legal Sphere	Geopolitical Sphere
	Human enhancement	Public/private ownership	Liability	Nano-divide
	Life extension			
		Business/University partnerships and conflict of interest	Business/University partnerships and conflict of interest	
	Cost and access	Cost and access		Cost and access
	Genetic discrimination	Genetic discrimination	Genetic discrimination	
		Intellectual property patents	Intellectual property patents	Intellectual property patents
		Economic "arms" race	Economic "arms" race	Economic "arms" race
Funding priorities	Funding priorities	Funding priorities		Funding priorities
Military applications	Military applications	Military applications		Military applications
Workforce preparation	Workforce preparation	Workforce preparation		Workforce preparation
	Privacy	Privacy	Privacy	Privacy
Technology transfer	Technology transfer	Technology transfer	Technology transfer	Technology transfer
Toxic waste	Toxic waste	Toxic waste	Toxic waste	Toxic waste
Non-biodegradability	Non-biodegradability	Non-biodegradability	Non-biodegradability	Non-biodegradability
Nanotoxicity	Nanotoxicity	Nanotoxicity	Nanotoxicity	Nanotoxicity
Nanocarcinogens	Nanocarcinogens	Nanocarcinogens	Nanocarcinogens	Nanocarcinogens
Worker safety	Worker safety	Worker safety	Worker safety	Worker safety
Definitions and standards	Definitions and standards	Definitions and standards	Definitions and standards	Definitions and standards
Nano-regulation	Nano-regulation	Nano-regulation	Nano-regulation	Nano-regulation
Public awareness and acceptance	Public awareness and acceptance	Public awareness and acceptance	Public awareness and acceptance	Public awareness and acceptance

itself is an exercise of social power that disenfranchises those with limited power.[10] For example, by placing workforce issues into the category of education, rather than the category of ethical issues, concerns related to economic disruptions and inequities are harder to surface. They become opportunity costs, which are primarily a threat to global competitiveness if not taken, rather than a direct harm to displaced workers and their families.

Sorting specific categories of concern by their moral dimensions forces a very different lens on each issue. I would add at least three additional moral dimensions to Lewentein's insightful list. The first is a general concern for the dignity and freedom of the human person. Enhancement technologies are a particularly good example of how nanotechnology may challenge the nature and meaning of the human person through advances in medicine and the convergence of technologies related to human cognition and genetic manipulation. However, questions of human dignity and freedom are also raised when education and other public relations efforts are used to coerce or manipulate public support either for or against emerging technologies. Likewise, the compulsory nature of technology tends to undermine human agency insofar as we are compelled to adopt or accept technologies that may not enhance our lives or develop our unique human potential. Such compulsion goes beyond the social power that drives policy and markets because it is inherent within the dynamic nature of technology itself.

The second moral dimension I would add is fidelity. Any issue that generates competing loyalties provokes an ethical dilemma. Examples of competing loyalties abound:

- Pressure from shareholders to increase profits despite consumer safety issues
- Global competitiveness at the expense of job security for workers
- Homeland security at the expense of individual privacy
- Publicly funded research generating privately held patents and proprietary methods
- Funding research in high-tech medical interventions when a large percentage of the population does not have access to basic preventive care

In these situations, individuals and institutions must choose between the competing demands of other parties, including personal and organizational conscience.

Finally, I would propose a dimension that might be termed material and existential harm. One could certainly argue that any ethical violation causes some level of harm, with harm being defined broadly as the result of a wrong act. An injustice is a wrongful act that results in harm, but justice is the root of the harm. In the dimension of material and existential harms, the harms need no other reference point other than harm itself. Material harm involves those tangible harms that can be objectively observed, measured, and compared. Chemical spills and toxic exposures are overt harms. They are most likely unintended and may have been unavoidable in retrospect, but they result in a direct and obvious harm nonetheless. They may raise questions of justice, fidelity, or power, but a portion of the harm also exists independent of these.

Existential harms involve situations in which a harmful act, or the failure to act in the face of harm, diminishes us all by undermining the basic human virtues that nurture and support human community. These harms represent an absence of will, a lack of courage, or a dearth of wisdom to do the right thing or seek the better alternative—but on a large scale. Society itself is complicit either by endorsement or lack of objection. Military nanotechnologies that reduce combatant casualties, or result in overwhelming force, may well make war a more palatable option for resolving political conflict, while failing to address the myriad other practical and moral objections to armed intervention. Embracing war as a matter of expediency or convenience on the grounds we cannot lose, flies in the face of experience and the goals of a peaceful and prosperous coexistence. Another example would be the failure to make available free or at low-cost certain nano-enabled technologies that raise the standard of living in the poorest countries. This might include technologies to purify water and mass produce vaccines for endemic diseases—conditions we know from experience to be barriers to the most basic improvements in the human condition for the most vulnerable among us. A final example might be continuing the pattern of stalling regulations on questionable environmental impacts of new technologies in favor of short-term industry profits, a near-sighted strategy that we know from experience simply piles problems and costs on to future generations—the children and grandchildren who have no choice but to trust that we will act in their best interests. Each of these examples represents a promised benefit of nanotechnology that could just as easily become a deeply seated harm, affecting all of us, at some level, possibly for generations, if it materializes technologically without a proper and balanced moral consideration.

In short, the moral dimensions listed in Table 3.2 move us away from impersonal social spheres and categories of concern, and back into the realm of human relationships and mutual accountabilities. Such an approach to classifying societal concerns also expands on Schummer's interest groups and meanings[4] as it encourages us to think across the boundaries of interest groups and in terms of commonly shared human values and preferences.

3.2.4   Pace, Complexity, and Uncertainty

In the end, how we come to identify and label "societal and ethical" implications will likely drive how we ultimately study and respond to them. Here again, pace, complexity, and uncertainty complicate the challenge. Nanoscience and technology are moving ahead rapidly on the technical

TABLE 3.2

Moral Dimensions of Societal Impacts
Dignity and freedom of the human person
Material and existential harm
Justice
Fairness and equity
Fidelity
Power and control

front, so the proverbial clock is ticking. Studying every possible issue across every possible sphere of impact, and then pulling that all together in a coherent strategy, seems unlikely in any reasonable timeframe. The boundaries between spheres are fuzzy and complex. The entire enterprise is rife with uncertainty. A more dynamic, systems-oriented approach will be needed in which we identify broad patterns of impact and ethical concerns that can help us anticipate, prioritize, and respond in timely and effective ways.

3.3   The Public Interface of Science and Human Values

A major topic of discussion and focus of funding in the United States has been garnering public acceptance of nanotechnology. This is interesting given the general support for science and technology that characterizes modern life. Gaskell and his colleagues (2005)[11] compiled the results of comparable social surveys in the United States, Canada, and the European Union to assess public opinion on who should make decisions on science policy and how such decisions should be guided. Nanotechnology, biotechnology, and information technology were the three areas of science policy targeted by the survey. In general, the researchers found a majority who favored current science policy and regulatory systems. Participants were, by and large, optimistic about technology and supported a scientific basis for decision-making. However, there was a substantial minority who were a bit more pessimistic about technology, and less confident in current scientific institutions. This group tended to favor decision-making based more heavily on moral and ethical considerations and more public engagement in the process. They were characterized by strong religious beliefs, lower educational attainment, and a general distrust of institutions.

Common sense tells us that the public is likely to value technologies they perceive enhance their lives in some way—convenience, entertainment, good health, secure employment—and resist technologies that overtly threaten

COMMENTARY

Rosalyn Berne

During the industrial revolution, several groups experimented with alternative ways of bringing technology into society. The "Shakers," for example, believed that everything they created should be a reflection of the inner spirit. While it's true that the Shakers didn't survive their ideology, they remain a remarkable example of how values can conscientiously be placed at the center of technological intent. Why not with nanotechnology, too? This "next technological revolution" has already become a marvel of human ingenuity and curiosity; a plethora of burgeoning technologies that point toward the exceedingly powerful increasing capacity of humans to alter, manipulate, and control matter with precision. The question is whose values are at its center. Will nanotechnology develop from an abiding care for humanity and reverence for the earth? Not unless its leaders embrace and articulate such intention, and its consumers, investors, and other stakeholders insist on it.

Every day, the world over, nanoscale research scientists and engineers refine and increase human knowledge and technological capacity. Their work, driven in large measure by the personal desire to alleviate suffering and enrich human life, could lead to the creation of processes, systems, consumer goods, and devices that will catapult human well-being and prosperity. Can this be accomplished with justice, and without sacrificing the health of the planet in the process of doing so? I believe so, yes; but only if all efforts are centered conscientiously in that intention, the value of which is embraced by all concerned.

Rosalyn Berne is Associate Professor of Science, Technology, and Society in the school of Engineering and Applied Science at the University of Virginia. Her primary research concern is the societal and ethical implications of nanotechnology, and its convergence with bio and information technologies, and cognitive sciences. In particular, she writes about the role of research scientists, human enhancement, and the use of technology in procreation. She is the author of *Nanotalk: Conversations with Research Scientists and Engineers about Ethics, Meaning, and Belief in the Development of Nanotechnology* (Lawrence Erlbaun Associates 2005), and has just completed her first nanoscience fictional novel, *Waiting in the Silence* (not published).

cherished values or security. The less we understand or recognize the immediate value of something, the more suspicious, resistant, and fearful we are
likely to be. However, Kulinowsky (2006)[12] points to a somewhat different
dynamic. She notes that new technologies tend to have strong public support
initially as they hear about potential benefits from technology advocates.
However, support then "tends to decline over time as values get articulated
and risks become the focus of debate."[12] She terms this moving from "wow"
to "yuck," often driven by media reports, public interest groups, and popular media, such as books and movies. Mehta and Hunt (2006)[13] affirm this
dynamic by considering the trajectory of earlier nuclear, information and
biotechnologies. They assert that technological developments roll out faster
than we are able collectively to understand and manage, in which case there
is a predictable "pattern of development, use, social concern, regulation, and
ultimately, some form of resolution."[13]

Implicit in this pattern is a certain time frame in which there is adequate
space for the process to work through. The high level of concern being
expressed about garnering public support at this point in nanotechnology's
development may be an effort to alter this pattern, or a recognition that the
pattern may alter itself. The speed with which nanotechnology may emerge,
in combination with possibly unprecedented levels of sociotechnical complexity and uncertainty, might preclude the necessary time needed to move
through the process. Alternatively, a significant event, such as a serious
product recall or highly controversial application, could spill over into public
support for nanotechnology in general.

It is at this interface between science and human values that social and
ethical implications research is poised. We turn now to one of the few models we have for potentially anticipating and minimizing controversy and
untoward outcomes of an emerging technology.

3.4   The Human Genome Project and ELSI

The Human Genome Project (HGP)[14] provides the precursor to the NNI's
efforts to examine societal implications. Formally begun in October of 1990,
the Human Genome Project was an ambitious and unprecedented scientific effort to unlock the mysteries of our human genetic composition and
function. A triumph of modern science, the overriding goals of the project
were to identify and map all of the human genes, as well as determine the
complete sequence of human DNA. If successful, the fully mapped human
genome would enable researchers to identify and understand the mechanisms of human health and disease. Just as the germ theory of disease in
the late 1800s ultimately led to the development of modern hygienic techniques, antibiotics, and antiviral medications, the human genome promised
to unlock the secrets of human genetic disorders and susceptibility to other

forms of disease. A detailed understanding of the structure and function of the human genome represented tremendous potential for solving some of the most challenging problems in modern medicine, including cancer, heart disease, neuromuscular disorders, and mental illness. In addition, a detailed knowledge of the structures and mechanisms of human genetics would eventually lay the foundation for personalized medicine. Rather than employing generic treatments that work for most patients most of the time, a personalized approach to medicine is one in which treatments can be efficiently and effectively targeted at the level of each individual person and his or her unique genetic and metabolic make-up. In fact, many supporters of the project viewed breakthroughs in human genetics as inevitable developments in our understanding of our human origins and functioning. The detailed mapping of the human genome itself was simply a logical and necessary step along the path to conquering human disease and relieving suffering on a broad scale.

However, the Human Genome Project was controversial from the start. Many scientists and policy makers believed the project was not technically feasible and would result in an enormous waste of valuable time and scarce resources. And, in fact, the criticism went deeper than just allocation of time and resources. Many were concerned with how genetic knowledge would be used or misused. Genetic discrimination was seen to represent a primary ethical threat if widespread genetic testing resulted in access to an individual's genetic profile by employers or health and life insurance companies. Moral questions were also raised about how deep knowledge of genetics might affect human self-perception. Religious leaders and others were uncomfortable with the suggestion that explanations for all human behavior and response to the environment could be reduced to simple mechanical interactions between and among genes. Finally, perhaps the greatest unease was the question of what potential harms could come from the ability to manipulate and control human genetics at its most basic level. Dire warnings were leveled about everything from designer babies to genetically engineered weapons of mass destruction. As it turned out, the technical success of the project exceeded everyone's expectations while the ethical challenges largely remain.

Predictably, many of the same general societal implications of the Human Genome Project have been leveled at nanotechnology. Categories of societal implications listed on the website of the Human Genome Project are instructive because one can readily pose the same basic topic areas in a survey of societal implications of nanotechnology. These include: fairness in the use of genetic information, privacy and confidentiality, psychological impact and stigmatization, reproductive issues, clinical issues and uncertainties regarding testing, conceptual and philosophical implications, health and environmental issues, and issues related to commercialization of products.[15]

In response to these critics, designers of the project established and funded a key research area that came to be known as ELSI, or ethical, legal, and social issues research. The U.S. Department of Energy and the National Institutes of Health devoted between 3% and 5% of their total annual HGP budget to the study of ethical, legal, and social issues associated with the creation and availability of genetic information and its associated technologies.

The results of this effort are mixed (ELSI Research Planning and Evaluation Group, 2000).[16] A 2000 evaluation of the program described major accomplishments in all four target areas of privacy and fair use (policy recommendations), clinical integration (state-of-the art genetic counseling), genetic testing (informed consent), and education and resources.[16] Supporters point out that the ELSI model is now applied around the world with respect to a variety of scientific endeavors. Research within the Human Genome Project continues and the ELSI website of the Human Genome Project has become a clearinghouse for current research and news.[15] Detractors of the effort contend that we still lack a coherent public policy on the range of controversial issues raised by the Human Genome Project, and ELSI research has had little impact on actual policy (Fisher 2005; Ramsay 2001),[17,18] despite the rapidly growing genetic industry. In either case, advances in genetics are rapidly evolving and fears about designer babies, medical privacy, patenting of genes and other life forms, and commercialization are materializing more or less unchecked. Nonetheless, the NNI took a page from the Human Genome Project's book in establishing its own version of ELSI research.

3.5   The National Nanotechnology Initiative and SEIN

Following the precedent set by the Human Genome Project, and in anticipation of the enormous potential impact of nanoscience and nanotechnology, the National Science Foundation (NSF) sponsored a workshop in 2001. The subsequent report, *Societal Implications of Nanoscience and Nanotechnology* (Roco and Banbridge 2001),[1] is a collection of expert opinions and recommendations on how best to "(a) accelerate the beneficial use of nanotechnology while diminishing the risks; (b) improve research and education; and (c) guide the contributions of key organizations."[1]

The first recommendation reads as follows: "Make support for social and economic research studies on nanotechnology a high priority."[1] The recommendation goes on to specify the inclusion of social science research within the nanotechnology research centers and suggests creating a distributed research center for social and economic research. There is an explicitly stated intent to incorporate the means for openness, disclosure, and public participation into the heart of the research and development program for nanotechnology. Additional recommendations address the need to inform, educate, and involve the public; create an infrastructure for interdisciplinary evaluation of the scientific, technological, and social impacts in the short-, medium- and long-terms; and educate a workforce.

The National Nanotechnology Initiative at Five Years: Assessment and Recommendations of the National Nanotechnology Advisory Panel (President's Council of Advisors on Science and Technology 2005),[19] is the first major progress report on the National Nanotechnology Initiative (NNI) released by the President's Council of Advisors on Science and Technology. As noted in an earlier chapter, with respect to societal concerns, 8% of the total NNI budget for 2006, or about $82 million, was earmarked for societal implications activities, sometimes referred to as SEIN (societal and ethical implications of nanotechnology). Approximately $38.5 million was targeted for programs working on environmental, health, and safety research and development. The remaining $42.6 million was split between educational activities aimed at workforce development, promoting public understanding and acceptance, and research on the broad implications of nanotechnology for society, including economic impacts, barriers to adoption, and ethical issues, particularly as related to research priorities (Nanoscale Science, Engineering and Technology Subcommittee 2005).[19] All research projects funded by the NNI are required to include a societal dimensions component that addresses one or more of the above categories.[19]

Presently, the Societal Dimensions Program Component Area (PCA) of the NNI's Strategic Plan (NSET 2004)[20] encourages and funds research initiatives in three broad areas: 1) environmental, health, and safety (EHS) impacts of nanotechnology development and risk assessment of such impacts; 2) education-related activities, such as the development of materials for schools and universities, as well as public outreach; and 3) identification and quantification of the broad implications for society. Social, economic, workforce, educational, ethical, and legal implications are included in the third category.[20]

In 2007, there was a shift in funding with approximately $44 million earmarked for EHS programs and $38 million for the combined areas of education and all other societal implications (NNI 2006).[21] Among SEIN researchers and other observers, the EHS allocation is somewhat controversial. For example, one might expect environmental, health, and safety concerns to be part of the basic cost of research and development funding as it is in general industry. Critics suggest this assignment of EHS to societal implications effectively reduces the budget for all other societal implications research to far less than what is needed, making SEIN research commitments largely symbolic, geared more toward managing public perception and support than minimizing negative impacts (Berube 2006).[8] There is some substance to this concern if one considers that in the proposed budget for 2008 (NNI n.d.),[22] the EHS budget rises to $58.6 million out of the $97.5 million earmarked for societal dimensions. Therefore, the amount of funding targeted at all other societal dimensions research will remain the same, with the relative percentage dedicated to non-EHS research shrinking from 46% to 40%.[22] In addition, while societal dimensions commanded 8% of the original NNI budget, this has now dropped to 6.5% despite strong calls for large increases, especially to the EHS portion of the budget.[22]

Funding aside, perhaps most telling is the single comment about non-EHS activities in the *FY 2008 Budget and Highlights*.[22] "Societal aspects of nano-technology are being addressed through several approaches, and agency-sponsored education and public engagement activities are reaching across the country." There is passing mention of workshops sponsored by NSET on public participation in nanotechnology and ethical aspects; however, the only items highlighted by name are the websites of the Center for Learning and Teaching at the Nanoscale, the Nanoscale Informal Science Education Network, and the traveling exhibit "Too Small to See," which was visited by over 350,000 children in 90 days at Walt Disney World's Epcot Center. Granted, public awareness and understanding of the science is crucial to an informed societal dialogue, but there is a clear priority in the societal agenda for NNI funding to garner public acceptance. Critics have been quick to charge that this focus is misplaced and will likely result in the NNI failing "to achieve its stated goal of maximizing the social good and minimizing the negative aspects of the nanotechnology revolution" (Sandler and Kay 2006), because it fails to integrate social and ethical implications research, and public engagement into the actual process of making meaningful decisions about the goals, priorities, funding, and constraints.[23]

If played out as predicted, and such a failure proves to be by intention and design, then it will also be an unfortunate ethical failure of both loyalty and truthfulness, as well as an abuse of power. If simply mishandled, the SEIN component will represent a number of missed opportunities. However, if done well and with good intention, the integration of SEIN research and public involvement into the fabric of science and technology research and development could introduce a more robust and adaptive paradigm for dealing with the challenges of rapid pace, complexity, and uncertainty. Time will tell whether this aspect of the NNI is more or less successful than that of the Human Genome Project in guiding policy and other responses to the broad societal implications of nanotechnology and the ethical questions it raises.

3.6   Origins of the Precautionary Principle

Initiatives similar to the NNI have been launched in many countries, with varying approaches to societal implications. If the United States tends to push ahead with science and technology, sorting out the effects as we go along or at some later point, Europe, Canada, and many other countries have moved in a more cautious direction when considering policy related to societal impacts. With its roots in the environmental movement in the 1970s, the Precautionary Principle has emerged as a strategy to cope with risk and scientific uncertainty (COMEST 2005).[24] It has been applied to treaties and international declarations on sustainable development, environmental protection, health, trade, and food safety.

There are many versions and interpretations of the Precautionary Principle, and we will return to a more detailed discussion in later chapters; however, the basic concept is relatively straightforward. Sometimes classified as the "better safe than sorry approach," the principle suggests that under certain conditions, anticipatory actions must be taken to avoid or diminish harm, up to and including, refraining from research and development activities altogether. Such conditions include the threat of potentially unacceptable harms in the face of considerable scientific uncertainty. This approach differs from earlier environmental approaches that identify harms once they have occurred and then attempt to mitigate those existing harms and prevent future ones via fines and other regulatory penalties and interventions.

Although it sounds straightforward, the principle itself is quite controversial in its interpretation and application. There is no general agreement on the definition of a harm, the magnitude of either the harm or the level of uncertainty needed to trigger a particular level of precaution, or what counts as a precautionary measure (Gardiner 2006).[25] Critics are quick to charge that it is unscientific and overly risk adverse. It is felt to unnecessarily impede the progress of beneficial research and development through wasted time and resources (Resnick 2003).[26] Proponents, on the other hand, argue that it is a reasonable response to plausible threats, especially in the face of increasing complexity and uncertainty. They cite examples, such as global warming, in which scientific uncertainty was used to justify inaction rather than address the predicted risks through effective policy initiatives (Ascher 2004).[27]

The implications of nanotechnology clearly go beyond immediate and future environmental threats, raising legitimate concerns that an overly stringent application of the precautionary principle will stifle the emerging science and delay the commercialization of valuable applications across the full array of industry sectors.[13] Because so much is unknown about the properties of materials at the nano-scale, the level of uncertainty in many applications will be high. Much of the science simply does not exist yet, and is likely to emerge in unpredictable ways across scientific disciplines and industries. In addition, the sheer scope of societal implications makes the precautionary approach feel overly simplistic. There are also risks and costs associated with not proceeding with a promising line of research. At the same time, the specter of prior failures of caution, despite early warnings, suggests unsettling analogies. Asbestos, radiation, PCBs, halocarbons, and others are all examples of our tendency to ignore or suppress evidence and delay needed action in the face of harm (Harremoes et al., 2001). Is there a tipping point at which public confidence in science and technology may force the adoption of some form of precautionary approach?

3.7 The Citizen as Moral Agent

Democracies work, in part, on the theory that a reasonably informed public is able to evaluate and elect the most suitable persons to represent them and their interests within the bodies of government. The various branches of government then use the resources and power granted them to serve the public good. One of the keys to making this all work is the phrase "a reasonably informed public." As we have seen, the relationship between science, business, and governance is a complicated mix of interests, goals, values, and assumptions, that may or may not always align with public values and perceived interests. Not surprisingly, science and technology policy is one area in which a reasonably informed public may not exist, complicating the role of the public in the policy process.

On average, most Americans know very little about basic science facts or the scientific process, and this situation is not improving over time (National Science Board 2006).[29] An often cited example is the fact that less than half of the American population accepts the theory of evolution despite strong evidence and widespread acceptance in the scientific community. Nonetheless, most Americans share a favorable attitude toward science and they support government funding of research. They also trust the leadership of the scientific community more than any other profession, except the military.[29] In fact, in research cited by the National Science Board, 84% of Americans agree that the benefits of scientific research outweigh the harmful outcomes as compared with only 52% and 40%, respectively, of their European and Japanese counterparts. Furthermore, survey research from 2004 reports that 83% of Americans agree, "even if it brings no immediate benefits, scientific research that advances the frontiers of knowledge is necessary and should be supported by the federal government."[29] However, this generally favorable view does not mean their support is totally unqualified.

Again, in research cited by the National Science Board,[29] more than half of the respondents expressed concern that we depend too much on science as opposed to faith; that science does not pay enough attention to society's moral values; and that scientific research has created at least as any problems as it has solved. For example, public concerns with the need to protect the environment are increasing, most recently with the media spotlight on global warming. However, if the public is to support an informed national science policy, then they themselves must have at least a basic knowledge of science and an awareness of key developments. This may be particularly true of nanotechnology given the scope of its potential impact, and the possible scope of a backlash should there be untoward outcomes. However, research previously cited suggests that the public may be only minimally and incompletely aware of nanotechnology and its implications.

We have previously explored the moral obligations of scientists, engineers, corporations, and policy leaders. We leave this chapter with a comment on the moral obligations of the citizen. Societal implications, by definition, involve potential effects on citizens, individually and collectively. We can argue persuasively that scientists, engineers, corporations, and policy makers are obligated to protect the public from avoidable harm, but the public also has a certain obligation to protect itself. We can argue in favor of obligations to inform the public of risks, respect their values, and invite their consent and participation; however, they must make the effort to be informed; assess the risks; reflect critically on their own values in light of a common social good; understand what they are or are not consenting to; and be willing to participate directly or indirectly in the process.

Paternalistic approaches to assessing and responding to societal implications that treat the public as little more than uninformed children or barriers to progress, undermine their essential dignity and underestimate their value and potential power. More important though, paternalistic attitudes and processes on the part of science, business, and government also encourage the public to act like uninformed children. This undermines society's ability to respond to change in positive and productive ways. The paradigm in which an informed and engaged public is welcomed into setting goals and priorities for science and technology, allocating resources, evaluating controversial directions, and identifying prudent constraints does not currently exist. I suggest that this reality has far less to do with an evil conspiracy in the halls of power, and more to do with the difficulty the average citizen has in choosing what social issues warrant his or her attention, and then finding an effective avenue through which to contribute. Not everyone is going to sign up to testify before Congress or serve on citizen advisory or oversight boards. Most interested persons will simply write a letter to their representative, contribute to an NGO that represents their interests, target their votes come election time, or talk to neighbors and friends. However, these are all useful actions that so promote the moral agency of citizens and the health of a democratic community. As with any issue, only a small, interested, and motivated group of citizens will select nanotechnology as their focus. The key to enabling meaningful public engagement will be the creation of paths of communication and impact that capture their attention and enable these persons to participate.

3.8 Questions for Thought

1. How would you set about describing or classifying the societal implications of nanotechnology?

2. Of the various societal dimensions and issues discussion in this chapter, which, if any, do you find troublesome? If you had $100 million to work with, how would you allocate the NNI budget for societal implications?

4

The Language of Ethics

4.1 Speaking the Language of Ethics

Science is organized knowledge. Wisdom is organized life.

—Immanuel Kant

The term **ethics** regularly tends to get tacked on to societal implications research, yet it gets very little direct attention. Although often mentioned in passing in articles and presentations on societal implications, the discussion is apt to either lack depth or swing to the academic extreme of being too conceptually dense to be of use in normal discussion. A language that is impenetrable to the average person is simply not likely to get used much. Yet, we do speak the language of ethics on a daily basis. Most of us simply don't realize we are doing so.

Whether referring to ELSI (ethical, legal, and societal implications) or SEIN (societal and ethical implications of nanotechnology), the term "ethics" seems to imply there is a set of implications that are separate from other legal, cultural, political, economic, and societal implications in general. Granted, this image is reinforced by the scholarly approach of many philosophers and ethicists. By focusing their attention on highly abstract and largely inaccessible approaches to moral analysis and argument, they lose the audience. While there is great value in the scholarly approach to ethics and morality, it often promotes an overly narrow and intellectually elite conception of ethics. For our purposes, we want the best of both the conceptual approach to ethics and the applied aspect of ethics that is present in the everyday lives of individuals and organizations.

In reality, few of us get out of bed in the morning thinking, "How can I do harm in the world today?" or "What unethical action can I take to get the day started?" By the same token, if we arrive at the end of the day with a knot in our stomach and a general feeling of discomfort, one can almost guarantee there is a moral element that has been violated at some point during the course of the day. In some small way we have been disloyal or dishonest; we've been unjust or failed to avoid a needless harm; we've failed to act to benefit another; or, we have sought short-term gain while knowing the longer term consequences of our action will be negative.

Human beings are, by nature, also moral beings. We are inherently social and our ability to live in community is enabled, in part, by our willingness to honor a set of shared values and precepts for how to treat one another. We implicitly agree there is a need to evaluate the positive or negative value of our actions. And, in fact, we do so all the time. In some cases, we codify those values in the law and then refer to them in the language of the law. In most cases, we use a lay dialect of the language of ethics. The terms ought and should, right and wrong, good and bad, fair and unfair, harmful and beneficial all reflect judgments that are both practical and ethical in their nature. Other terms are a bit more problematic. For example, to say I have a right to something may reflect a clear legal entitlement, but not necessarily an undisputed moral right. It is, in part, the lack of clarity and depth in how we use this more casual dialect that can render discussion on moral issues uncomfortable and contentious.

Consider the ongoing debate on abortion in the United States. Language has come to be used far more as a weapon than a tool for dialogue. In fact, the abiding circularity and ineffectiveness of this particular debate is a tribute to the damage that can be done when words and language are used poorly. The categories "pro-choice" and "pro-life" are themselves meaningless misnomers that oversimplify the debate in ways that inflame rather than inform. Advocates for abortion are not opposed to life per se. Most opponents of abortion are not advocating absolutely no choice when it comes to reproduction. The two camps end up literally speaking two different vocabularies within the language of ethics and failing to find the common ground upon which they can build a solution. I imagine all but the most virulent and impractical voices on either side would agree that, as a matter of the common good, it is morally preferable that babies are loved and provided for. Likewise, most everyone would agree it is morally preferable that women (along with their partner and any existing children) are treated with the dignity implied by basic human freedoms of thought, conscience, opinion, and expression (United Nations 1948).[1] Most of us would also agree that it accomplishes little to punish or threaten a woman economically, socially, culturally, and politically if she chooses to carry a child to term. In other words, rather than right to life and right to choice questions, a more useful ethical question becomes what WE, as a society, owe both a pregnant woman and the embryo or fetus she carries. Agreeing upon these basic moral premises would do much to yank both sides out of the constitutional courtroom and into a discussion of the practical realities of unplanned, unwanted, unprepared, or medically precarious pregnancies, and how best to either avoid them in the first place or provide an adequately compassionate and supportive response to each woman's unique situation, needs, and abilities. The entire tone, direction, and focus of the debate is altered for the better in a matter of a few differently chosen words, leaving much for us to agree upon in moving forward on an issue of tremendous moral complexity.

For our purposes, the combination of theoretical and applied ethics can be a highly practical and effective means of identifying and sorting through everything from the most philosophically profound issues to the most mundane, everyday, pragmatic concerns associated with science and technology. Ethics is best used as both a language and a framework. This language can be used to define a limited set of ethical considerations that can then serve as the basis for a common dialogue. Within such dialogue, societal concerns can be integrated with the objective realities and potentials of emerging science and technology.

In this chapter, I will provide a brief conceptual presentation of ethics in order to introduce the language of ethics to be employed in Chapter 5, and then used throughout the applied chapters in Parts Two and Three of the book.

4.2   Ethics 101

Ethics represents a branch of philosophy concerned with a systematic understanding of the moral basis of right action. When faced with a difficult choice between actions, ethical reflection is a source of guidance. An ethical problem occurs in a situation in which there are moral factors to be considered in selecting an action. An ethical dilemma involves a situation in which all choices have both good and bad or desirable and undesirable effects. When uncertain about whether such a dilemma even exists, the language of ethics offers a variety of lenses through which to identify whether or not a moral dilemma is present. Ideally, ethical or moral reasoning is a routine step in any decision making process. In fact, as I keep reiterating, you engage in moral reasoning on a regular basis whether you realize it or not.

For example, deciding what to have for dinner may not appear to have a moral component. However, if you are a committed vegetarian faced with the same choice, the question is not as easy. You'll need to decide whether to avoid just meat or also eggs and dairy products. What about the many processed foods that contain animal byproducts? The answers to these apparently simple questions are embedded in a complex matrix of values, beliefs, and moral judgments, many of which may not even be consciously apparent in your consideration.

Moving a bit closer to technology, choosing between a hybrid car and an SUV is also a moral choice as well as a practical one. If you live 25 miles from the nearest town at 10,000 feet in the Colorado Rockies, then you may have a much stronger utilitarian justification for your purchase of a gas guzzling SUV than someone living in Los Angeles who simply prefers driving a larger and more substantial vehicle. Reporting all income properly on your income or business taxes is a moral choice. It involves your perceived duty to contribute to whatever common good is derived from taxation, including the tax-funded interstate highway system or public schools. The decision by a corporation to withdraw a product from the market for safety reasons is a prudent choice for utilitarian legal reasons, but exactly how and when the corporation does it is also a reflection of the ethical principles of veracity (truth-telling) and fidelity (loyalty) to the consumer.

The art and the science of ethics lie in whether or not and how we recognize those questions that call us to a deeper and more profound consideration. As such, ethics is simply a great method of critical thinking. Asking too few or the wrong questions nearly always gets you to an incomplete or ineffective solution. Failing to ask the questions at all is, first and foremost, a failure of critical thought. On the other hand, routinely asking good questions is a strong, disciplined practice of effective and responsible thought. It is also a form of moral leadership no matter your position in relation to the situation at hand. Finally, the integration of moral analysis into basic personal and professional decision making offers a solid strategic advantage as governmental and business entities come under increased public scrutiny and accountability.

At this point, it may be helpful to distinguish between two significant branches of ethics. **Metaethics** is the more abstract and conceptual approach that examines the very nature of moral beliefs by answering questions about the source of morality and the meaning of basic concepts such as right or wrong. This is where the formal language of ethics originates, and the primary goal is to provide a logical and consistent foundation for moral deliberation. **Normative ethics**, on the other hand, is more oriented to the application of practical judgments to everyday actions by combining underlying ethical assumptions with factual information, commonly held beliefs and values, and socially and culturally accepted standards of behavior. The primary goal is to answer the practical question of what ought I or we to do in this particular situation, or in all similar situations.

There are a few other theoretical concepts that come in handy in ethics. These concepts help pinpoint your ethical comfort zone—the way you or your organization tend to react when confronted with a moral dilemma. They also help you recognize the comfort zones of others so that you can be more effective when facilitating a discussion in which moral positions are in conflict. For example, most of us fall somewhere along the continuum between **objectivism** and **relativism** in our general worldview. Objectivists assume there is an absolute source of moral truth, truth with a big "T" if you will. There is one best and right answer to any situation whether dictated by religious belief or a dedication to objective, rational analysis as advocated by many theologians and classical western philosophers. This is a comfortable worldview in that we can rely on a more or less consistent answer to moral questions and dilemmas. Relativists, on the other hand, are suspicious of black or white positions that fail to take into account the particular context of a decision or that fail to account adequately for cultural differences, diverse individual preferences, situational variation, or changing circumstances over time. While most of us tend to exhibit a bit of both types of thinking, people who tend towards the opposing side of the continuum from us are often the ones with whom we have the most conflict when trying to discuss

BOX 4.1 COMPARISON OF ETHICAL THEORIES

Theory	Major Assumption	Key Concepts
Utilitarian Ethics	An action is deemed morally acceptable because it produces the greatest balance of good over harm taking into account all individuals affected.	No action is in itself right or wrong apart from its consequences. The principle of utility determines that actions are right insofar as they produce the greatest happiness for the greatest number of people. Act utility holds that the principle of utility be applied to particular acts in particular circumstances. Rule utility holds that the principle of utility be applied to test rules that can then be used to determine the rightness of an action in similar situations.
Duty-based Ethics	Actions should be judged right or wrong based upon inherent characteristics (duties) rather than consequences.	Moral rules are objective truths that may or may not be absolute. A moral duty is an imperative that compels us to act or refrain from acting on the basis of duty alone. When duties conflict, one must act according to the most stringent obligation. Consequences may be considered secondarily to achieve the greatest balance of rightness over wrongness.
Virtue Ethics	An action is deemed morally acceptable because it seeks the "good" and serves the "good life," and because it is taken for the sake of virtue (as opposed to the sake of duty).	Virtues are natural inclinations that, when cultivated and shaped by education, experience, reflection, and proper intent, increase our happiness and enrich the quality and dignity of our lives. Moral character rather than moral principles is what is needed for morally sound decisions and actions. Virtues include benevolence, loyalty, compassion, honesty, charity, sincerity, respect, kindness, fairness, rationality, patience, perseverance, prudence, and courage.
Rights-Based Ethics	Natural rights are a gift of nature or God that cannot be taken away. Moral and legal rights are justifiable claims to which individuals and institutions can be held.	Universal human rights form the basis for establishing and/or evaluating ethical standards within the social order and include rights to life, freedom, political participation, legal protections, and basic social and economic goods.
Feminist Theory	Patterns of oppression in the political, economic, legal, and social spheres persist, and fail to address resulting injustices.	A continued imbalance in social and economic power among disadvantaged groups (women, the aged, children, the poor, ethnic and racial minorities, the disabled) is the primary basis for moral analysis. Strongly supports conceptions of autonomy, respect for persons, and justice.
Communitarian Ethics	The rightness of actions can be judged against an ideal of the common good.	Society is a complex web of intersecting communities that must dynamically negotiate a set of shared values that are aligned and accountable to the larger society. The common good is achieved through a balance of individual and societal responsibilities including individual liberty, sustainability, and intergenerational justice and civic engagement.

issues with a strong moral component. The same tendency holds true for stakeholder groups when dealing with issues on a larger scale.

Other common examples of ethical comfort zones reflect the underlying assumptions of various ethical theories and principles (see Box 4.1). As we briefly explore each perspective, consider an example of a decision you or your organization might have made that clearly used this perspective either explicitly or implicitly in its justification or rationale.

4.3   An Ethic of Utility

People whose moral foundation is strongly **utilitarian** instinctively perform a cost/benefit analysis in their heads when faced with a dilemma. Their primary concern is with the consequences of their actions. Derived from the work of consequentialist philosophers Jeremy Bentham and John Stuart Mill, the goal is often under-interpreted as simply acting to maximize happiness and minimize unhappiness. The term "happiness" is clearly problematic and might better be replaced with "the good" or perhaps "the greater good." An act is moral if it results in better or more desirable consequences than any other act in the same situation. Furthermore, no action in and of itself is considered good or bad apart from its consequences. Naturally, how one defines good, better, or desirable is open to interpretation and debate; however, we do generally aspire to a range of similar goods, both tangible and intangible. These might include love, friendship, knowledge, prosperity, security, and life itself.

Another common application of utilitarianism is found in the phrase "the greatest good for the greatest number." In this formulation of the concept of utility, one acts in the interests of the majority. The CEO of a corporation may be faced with the dilemma of cutting labor costs or losing the business and, by extension, putting everyone out of work. Although the people who lose their jobs or benefits are harmed, more people are likely to benefit if the company stays in business.

This perspective is perhaps the one most common in American popular culture and business. Strongly pragmatic in its application, it is also prevalent in science, economics, and politics. With respect to technology, there would ostensibly be little point in creating a technology that did more harm than good or that cost more to produce than it paid out in benefits. However, the devil is in the details. Most technologies have a dual nature as we have seen. We derive great benefits from our automobiles, but at an increasing cost to the environment, our health, and even national security if one considers the geopolitics of oil. When done effectively, utilitarian analysis forces a systematic consideration of the full range of consequences.

A strong critique of this perspective is that, when applied, we have a tendency to focus more on short-term than long-term consequences. Short-term consequences are what we will face immediately, and tend to be more amenable to prediction. However, to counter this tendency, the **slippery slope argument** recognizes that what seems acceptable in the short-term can slip out of control past our expectations, or simply have a different and less desirable outcome in the longer term. This argument is highly applicable to new technologies. Consider nuclear technology and the speed with which the atomic bomb moved from the weapon that ended a war to the weapon that initiated the Cold War. Nuclear technology continues to haunt us in today's context of political instability and terrorist aspirations, and perhaps poses more danger than it ever has in the past. Biotechnology is also susceptible to slippery slope arguments as we consider possible untoward and negative consequences of genetic engineering on the environment over time.

Another related critique of utilitarianism lies in the implicit assumption that "the end justifies the means." If the interests of the few can be sacrificed in order to derive the greatest benefit for the greatest number, and no action is inherently wrong, then certain individuals or classes of individuals are always at risk. With respect to new technologies, among the most vulnerable populations are the poor or the semi-skilled. These groups are the most likely to face displacement from newly automated jobs or be unable to find a job due to the need for a certain level of technological competence. The end is that work is accomplished more efficiently and the company prospers, but the means includes the harm to those least able to respond effectively to a change in their employment situation.

A final note on utilitarianism takes us back to Chapter 1 and the complicating factors of pace, complexity, and uncertainty. The more quickly one must make a decision, the harder it is to consider the full range of consequences. So, while utilitarianism acknowledges the need to account for complexity by considering a range of possible outcomes, it is often challenged to do so. Likewise, it goes without saying that if the outcomes are uncertain, we will be unable to make decisions solely on the basis of consequences. Utilitarianism is inherently limited by what we do not know and cannot anticipate with any certainty. When considering the potential implications of nanotechnology, a common argument is that the technology and its applications are too new to even predict the consequences. However, utility is not the only moral measure via which ethical dimensions can be approached.

4.4   An Ethic of Duty

Deontology, or **duty-based ethics,** involves a basic assumption that right actions are defined by our intention or motive and the inherent rightness or wrongness of the action itself, regardless of the consequences. This approach to ethics suggests that we have certain duties to one another that can't be dictated by the consequences of our actions alone, particularly when the likely consequences are hard to predict or largely unknown. For example, while a utilitarian will judge the relative value of telling a lie in each particular situation, a deontologist might refuse to lie in all situations on the grounds that lying is inherently wrong. In other words, we have a duty to tell the truth because telling the truth is the right thing to do. And there are valid reasons to support this assumption about truth-telling as a moral obligation.

People need the truth to make informed decisions. Relationships of trust are compromised when we can't be sure that others are being honest with us. Take, for instance, informed consent in research. Informed consent is based, in part, on the belief that research subjects in an experiment or study should understand the possible risks and benefits of the research before consenting to participate. Likewise, the concept of transparency in business is founded on the assumption that management has a practical and moral obligation to be truthful about specific aspects of the business' performance. By extension, transparency is an issue with emerging technologies insofar as the technology may have inherent risks related to worker or consumer safety, or negative environmental impacts of which we should be aware. Failure to inform leads to public backlash in the form of class action lawsuits, consumer boycotts, and stricter regulation.

Another common example of duty is the duty to avoid harm. There is no moral justification for causing avoidable harm for harm's own sake. This is an example of a particular ethical construct of classical Kantian ethics that is sometimes termed **universalizability**. It suggests that one should act in ways that can be universalized to all similar situations. Since it would never make moral or practical sense to desire or cause needless harm, one can universalize this to say we have a perfect duty to avoid needless harm in any situation.

However, we are again faced with a problem of definition. How does one define harm, and particularly in the face of multiple harms? Consider technologies used for advanced life support. These technologies are extraordinarily invasive and often do not lead to full recovery or even partial recovery of the patient. In addition, they are very expensive. I may define life "attached to a machine" as the greater harm while my children, in making decisions on my behalf, may judge death to be the greater harm. A third-party observer may define the greater harm as the resources wasted on expensive technologies and treatments that were not likely to succeed in the first place. Simply identifying a duty to avoid harm is not sufficient grounds for action. One must still weigh the moral context of the duty itself and define the language of duty accordingly. The definition of harm is open to interpretation and negotiation, thus the need for deeper moral reflection and dialogue.

Finally, duties often conflict with one another making it necessary to weigh one duty against one or more other duties. A company has a duty to maximize profit to its shareholders, which may conflict with its duty to protect consumers or the environment. The duty to obey the law may result in complying with poorly conceived regulations that prevent one set of harms while causing another set. It is common in these circumstances to see utilitarian analysis slip into the mix as parties assess the likely consequences of falling short on one duty versus another.

With respect to pace, complexity, and uncertainty, duty-based ethics are a mixed bag. Having a strong perception of duty and a sense of how to prioritize competing duties, irrespective of the consequences, can allow one to make decisions somewhat more quickly and consistently when time is short, the situation is complex, and the outcomes are uncertain. However, when the situation is complex, it is also not uncommon to have two or more duties in conflict. To complicate matters further, uncertainty makes it difficult to use consequences as a secondary measure when attempting to sort out the higher order duties.

4.5 An Ethic of Virtue

The roots of **virtue ethics** date back to ancient Greece and the idea that moral action originates in the character of an individual. Character develops over time and is shaped by both the natural inclinations of a person and the combined influences of family, culture, education, experience, and moral reflection. A moral act is one that promotes the good with the intent of the individual to do good. Motive is as important as outcome. The good is itself defined in terms of human excellence and thriving. A "good life" is one that allows us to achieve a level of personal happiness, as well as to serve the communal best interest.

A central premise is that individuals with a strong, well-developed character do not require rules and principles to guide their actions. Instead, they use their own moral intuition to call into action the necessary virtues to guide their action. The statement "I know the difference between right and wrong" is essentially a self-assessment of one's own moral intuition and ability to act on that intuition. In fact, I hear this frequently from students who resent having to take an ethics course as part of their professional training. By the end of the course, most of them admit to being humbled by the complexity of the issues and their struggles with being able to develop a coherent response for themselves. It is never a matter of them not knowing the difference between right and wrong at all. Rather, their ability to apply their moral intuition is incomplete when confronted with complex and unfamiliar situations.

Virtues are character traits that enable a person with the proper intent to do the right thing when faced with a difficult choice. The **moral virtues** typically include temperance, courage, love, justice, and dignity, along with honesty, loyalty, sympathy, and fairness, among others. **Practical virtues** have also been proposed to include such characteristics as patience and prudence. Practical virtues are morally neutral, but can enable virtuous behavior. For example, a prudent person will exercise a great deal of common sense when evaluating the proper virtuous response. While moral virtues are inherently moral in all situations, common sense can also be used with ill intent to serve non-virtuous ends. Thus, it is a practical or non-moral virtue.

In the professions, many codes of ethics rely heavily on a virtue orientation. We hold strong expectations that people in positions of power, such as scientists, engineers, physicians, lawyers, CEOs, and elected leaders, have an appropriately developed character and will act with virtuous intent. Clearly, this is not always the case and, although most of us do try to do the right thing, our moral intuitions can simply be wrong or, at least, incomplete. Since character develops over time and with experience, a decision I make today may not have been the one I would have made back in my 20s. In addition, character develops with reflection, which itself requires practice. Human beings are masters at convincing themselves their actions are justified, their motives pure, and their rewards deserving. Even at our best, we can be inconsistent in our judgments by failing to fully appreciate our own motivations or by lacking the experience to assess what is required in any particular situation. In reality, we may not always act on our moral intuitions because of fear, greed, confusion, lack of experience, or even simple convenience. Having the moral humility to recognize this potential failing is a step toward a more open and productive moral dialogue.

In assessing technology, questions regarding how this technology is likely to serve the purposes of human excellence and thriving force us to also consider our personal motives and intentions more deeply. Likewise, it forces examination of the intentions of our organizations and institutions. The virtuous person or organization routinely acts to balance self-interest with the greater communal interest, and does so with courage and clarity of intent. It is partly the failure of so many people in positions of power to act with character that has brought public mistrust of our organizations and institutions to such low levels.

As might be expected, this mistrust of power is amplified in the presence of increasing pace, complexity, and uncertainty. With respect to the assessment of technology, the faster the pace, the more complex the situation, and the more uncertain its outcome, the greater the need for individuals, organizations, institutions, and society as a whole to be able to work together. Assessing the meaning of a particular technology to human thriving requires open sharing of information and trust, followed by the shared character to either guide the technology's development or reject it.

4.6 Companion Theories

There are many other approaches to ethics. All have some value as a particular lens for evaluating a situation. For example, **rights-based ethics** examines a situation from the standpoint of what each party is either entitled to or is obligated to with respect to other parties. The concept of rights is itself complicated with there being many different ways to define a right and many different levels of rights. There are legal rights, moral rights, and political rights—and one right does not necessarily lead to the other. There are also positive and negative rights. So, the word itself is widely used with little precision; however, it is an important concept in any societal deliberation because it essentially defines the limits of what we can expect from each other or owe one another and society as a whole. The many and complex issues around intellectual property and patents with emerging nanotechnologies are examples of how rights intersect with the process of both science and technological innovation in the marketplace.

The ethics derived from **feminist theory** provides an interesting lens related to justice. Feminist theory contends that society continues to be plagued with an imbalance of power so that the dominant groups (those with power) are able to oppress those groups with limited power in a cycle that cannot easily be broken. It is not hard to see how this concern might apply to many technology-related issues insofar as those with resources and influence are able to either promote or block certain technologies and the manner in which they are developed. At the same time, groups with less power have little recourse when there are harms, such as economic displacement or environmental damage. Likewise, funding sources hold tremendous sway over what science does and does not get funded, how that science is shared with other stakeholders, and often, whether or not the methods or technologies that are generated are subject to regulation. A goal of feminist theory is to create pathways through which abuses of power are minimized via power sharing and the building of relationships among stakeholders.

Communitarian ethics presents an interesting contrast to the general orientation of Western ethics toward the primacy of the individual. Communitarian ethics emphasizes a concept of the "common good." The common good is an ideal that reflects the shared values of a group, community, or society. These values are ideally negotiated on a dynamic, ongoing basis in an essentially democratic process of dialogue. For our purposes, a central tenet of this perspective is that the rights of the individual must be balanced against the needs of the community. In other words, being a member of a community entitles me to certain rights, but it also obligates me to certain responsibilities that may limit my choices and certain freedoms. For example, a society has a vested interest in its own survival. This long-term interest allows values of intergenerational justice and sustainability to dictate that individuals are not entitled to unlimited resources but rather must accept some limitations so that the larger society can meet the wider array of needs now and into the future. Avenues of research or technologies that benefit only a few, but at a high cost to the economy, environment, or other aspect of the social order, can be legitimately restricted in this viewpoint.

4.7   Ethical Principles as Practical Tools

Principlism is an approach to ethics based on weighing basic principles that then determine conduct. Avoiding needless harm (**nonmaleficence**), acting in the best interests of others (**beneficence**), equitability and fairness (**justice**), truth-telling (**veracity**), and loyalty (**fidelity**) are examples of common principles in bioethics that frequently conflict with one another in an ethical dilemma (see Box 4.2). Several principles may apply in any given situation. Take, for example, whistleblowing. If I am aware that my employer is engaged in some sort of fraud, such as misinformation regarding product safety or environmental impact, then I have to weigh telling the truth (veracity) against loyalty (fidelity) to my employer, regulators, the broader public, and myself. I have to also consider the harm being done that is otherwise avoidable (nonmaleficence), the nature and scope of any benefits of the action (beneficence), and whether the action may result in an unjust imbalance of benefits and burdens for the various parties affected (justice).

We all have certain principles we find harder to violate than others. For example, we may be willing to lie rather than betray the confidence of a colleague, or we may be willing to expose research subjects to a somewhat questionable level of risk in the short-term in order to realize the benefits of a promising line of research in the longer term. In situations in which you feel some unease with the direction of the decisions being made or the actions taken, if you sort down through the list of moral principles, you can always pinpoint the principle or combination of principles that are in conflict. It is the nature of a moral dilemma that one has to violate one or more principles in order to adhere to one or more others. With respect to new technologies, this ethical approach calls us to weigh potential harms and benefits against each other along with other principles that raise questions of veracity, loyalty, or justice.

It is very important to recognize the limitations of both individual principles and the assumptions of various ethical theories. The theories and principles offer alternative ways of identifying ethical issues. They don't tell you what to do, rather they tell you how to organize and weigh your thoughts. Ethical deliberation is basically a process of applying reasoned analysis to multiple dimensions of a situation with ethical concepts functioning to

---

BOX 4.2   COMMON ETHICAL PRINCIPLES

Nonmaleficence: Avoid needless harm.

Beneficence: Act in the best interests of others.

Justice: Act in ways that are equitable and fair. Achieve a balance of benefits and burdens.

Veracity: Tell the truth.

Fidelity: Be loyal. Keep promises.

help identify the various dimensions to be considered. While philosophers, generally, and ethicists, specifically, argue ad infinitum about the relative merits of each concept, the concepts themselves are no more than practical tools for posing reasonable questions and stimulating disciplined reflection. The questions raised by these varying theoretical perspectives are implicit in the practical issues and concerns that arise when considering any societal impact of a new or emerging technology. The ability to recognize the particular lens in use is imminently helpful because it allows you to shift your own language and perspective to better understand and respond to alternative viewpoints (Bennett-Woods 2005).[2]

Take for instance the current debate surrounding embryonic stem cell research. Those in favor of stem cell research are largely speaking the language of utilitarianism and beneficence as they argue for the potential benefits of medical therapies derived from a better understanding and ability to manipulate embryonic stem cells. They are focused on the consequences of the research for a greater good. Those opposed to stem cell research are largely speaking either the language of duty or rights. They may claim we have a duty to protect unborn persons, even at the embryonic stage, purely on the basis of respect for the dignity of all human life and potential human life. By extension, they may argue that embryos should be accorded a certain basic right to life, at least a right to existence without being subject to destruction as a means for research. They are focused on the dignity of the human person and the inherent wrongness of treating a potential human being solely as a means for someone else's ends. For all intents and purposes, the two sides are speaking different languages, so it becomes very difficult to establish any sort of basis for a balanced dialogue. In reality, there are compelling points on both sides that warrant a reasoned consideration, minus the oversimplification and exaggerated rhetoric that characterizes the issue in the media.

4.8 All Research Is Human Subjects Research

Perhaps the most widely recognized application of ethical concepts in science involves human subjects research in medicine and the behavioral sciences. Modern medicine has come a long way from its primitive roots in which patients were subjected to largely untested (at least by current scientific standards) folk remedies, bloodletting, and various cultural rituals. The introduction of modern sanitation, nutrition, and antibiotics has largely shifted the focus of medicine in the developed world from the control of infectious disease to complex management of acute trauma and non-infectious diseases, such as diabetes and cancer. However, nearly every major advance in medical science has come as a direct result of research on living human subjects. Such research can often be risky and most subjects never directly benefit from their participation. Abuses in human subjects research, such as the infamous Tuskegee study, in which 400 black men with syphilis were observed untreated for 40 years in order to document the natural progression of the disease (CDC n.d.),[3] have led to a set of commonly accepted procedural standards. These standards are based on ethical principles that safeguard the rights and well-being of patients.

Fascinating, right? But how does this relate to the ethical and societal implications of nanotechnology? Commentators in nanotechnology have actually drawn an analogy between the potential impacts of nanotechnology and human subjects research (Bennett-Woods and Fisher 2004; Sarewitz and Woodhouse 2003).[4,5] For example, when exploring the societal implications of nanotechnology, Sarewitz and Woodhouse describe the "unfolding revolution as a grand experiment—a clinical trial—that technologists are conducting on society."[5] The direct implication here is that the considerations and review processes used for clinical trials in human subjects research may also be applicable to the various uncertainties posed by nanotechnology. The factors of rapid pace and increased complexity add to the overall uncertainty as the depth and breadth of potential impact multiplies the variables that would need to be factored in to fully assess the likely outcomes. However, pace and complexity also add to the practical challenge of trying to expand these narrowly targeted principles and processes from the level of the individual to the larger society.

The three principles that underlie the ethical treatment of human subjects are formally established in a document entitled the *Belmont Report* (The National Commission for the Protection of Human Subjects of Biomedical and Behavioral Research 1979).[6] The first principle, **respect for persons**, is the basis for the concept of patient autonomy (self-determination or self-governance) and requires guidelines for obtaining informed consent and the protection of vulnerable populations, such as children. The second principle, **beneficence**, obligates researchers to avoid needless harm by minimizing any inherent risks while also attempting to maximize the potential benefits to subjects. **Justice**, the third principle, examines more broadly how likely the benefits and burdens of the research are to be fairly distributed. Using the Tuskegee study as an example, all three principles were violated. The subjects were misled into believing they were being treated and so never gave informed consent; they were needlessly harmed since a cure was known and available; and, they bore the full burden of the research with no benefits to themselves or even to science since the stages of syphilis were relatively well documented at the time of the study.

Although the principles outlined in the *Belmont Report* are narrowly directed at protecting the individual human subject, they can theoretically be expanded to consider broader societal impacts. This conceptual expansion might provide at least a partial framework within which ethical analysis and dialogue can occur when assessing nanotechnology and those related technologies it will enable.

4.9   An Ethical Framework for Technology Assessment

Nanotechnology, as with other disruptive technologies, raises an array of practical concerns that can be readily expressed in the language of ethics. When considering the moral dimensions of technology, ethical considerations may be complex, but they are fairly stable over time. The same basic concerns tend to be raised again and again, albeit with a slightly different focus relative to the specific situation at hand. The questions of risk, sustainability, justice, fairness, and power raised by nanotechnology are much the same questions traditionally raised in both bioethics and the general area of ethics and technology (Bennett-Woods 2007; Grunwald 2005; Lewenstein 2005).[7,8,9] Of interest is the fact that these same basic concerns align closely with the Belmont principles used in human subjects research if expanded in scope.

The principle of respect for persons can be expanded to include respect for communities, while the principles of beneficence and justice can likewise be respectively broadened to incorporate the common good and a broader conception of social justice (Bennett-Woods 2007; Bennett-Woods 2006; Bennett-Woods and Fisher 2004).[7,10,4] Taken together, these three principles strive to balance concerns at the levels of the individual, the local community and the global society (see Box 4.3).

The first principle, **respect for communities**, focuses on the extent and nature of public participation in decisions regarding an emerging technology. The term community is admittedly vague, referring to a range of stakeholders (including interest groups), to socioeconomic and sociocultural groups, to national, state, and local municipalities. A central issue within the general scope of this principle then becomes the opportunity and ability of the community to exercise an element of autonomy and choice with respect to science and technology. Related issues include privacy and other legal and regulatory matters. Finally, this principle is concerned with both the identification and representation of vulnerable communities within the larger community (Bennett-Woods and Fisher 2004).[4]

BOX 4.3 ETHICAL FRAMEWORK FOR TECHNOLOGY ASSESSMENT

Principle of Respect for Communities: Act in ways that respect the ability of communities to act as autonomous, self-governing agents.

Principle of the Common Good: Act in ways that respect shared values and promote the common good of communities.

Principle of Social Justice: Act in ways that maximize the just distribution of benefits and burdens within and among communities.

Source: Bennett-Woods, D. and E. Fisher. 2004. Nanotechnology and the IRB: Toward a New Paradigm for Analysis and Dialogue. Paper presented at the joint meeting of the European Association for the Study of Science and Technology, and Society for Social Studies of Science. Paris, France. 2004. Paper available online: http://www.csi.ensmp. fr/csi/4S/index.php

---

The **principle of the common good** extends the more narrowly defined definitions of risk/benefit, which tend to be limited to largely utilitarian technical and environmental assessments. One example of guidelines that are almost exclusively targeted toward minimizing specific technological risks and safety concerns are the guidelines developed by the Foresight Nanotech Institute (2006).[11] The Foresight Guidelines for Responsible Nanotechnology Development (draft version 6) focus primarily on potential safety issues, risk/benefit assessments, and regulatory approaches. While these are critical issues to be addressed, the principle of the common good also calls for the assessment of outcomes that either support or violate generally held social and cultural values. It also promotes a broader definition of societal risks and benefits in the short-, medium-, and long-terms (Bennett-Woods and Fisher 2004).[4]

Finally, the principle of justice advocated by the Belmont report can be expanded to the **principle of social justice**. The goal here is to acknowledge the inherent social, economic, and political disruption associated with any major technological advance. Experience and common sense tell us that some communities will benefit more than others. Technology does not diffuse uniformly into society. Those with economic or political resources will generally have the highest level of access initially. To the extent that access to this particular technology further increases their economic and political resources, the technological divide between segments of society likewise increases. Other communities may be directly harmed by environmental, political, economic, and other societal consequences. As mentioned previously, workforce disruptions will most likely be felt most acutely by those who already have the fewest options for employment. A principle of social justice calls us to consider how best we can minimize burdensome consequences for these communities or, at the very least, compensate them in ways that are fair and equitable (Bennett-Woods and Fisher 2004).[4]

Each of these principles will be explored more fully in Part Two. For now, it is enough to simply revisit the discussion in Chapter 3 of using moral dimensions as an approach to describing societal implications of nanotechnology. One can readily see the connections between the proposed ethical framework and the moral dimensions cited in the earlier discussion. The principle of respect for communities seeks to honor the basic dignity and freedom of human persons by allowing them a measure of informed choice as a community when assessing the impacts of nanotechnology. The principle of the common good concerns itself with broad considerations of material and existential harm. The principle of social justice seeks just, equitable, and fair access to technology, as well as appropriate compensation for its disruptive effects. All three principles assume certain obligations of fidelity between the promoters of science and technology and the community. Finally, each of the principles seeks to offset inherent imbalances of power by encouraging information sharing, participatory decision making, values-based assessments, and obligations of protection and compensation for both harmful untoward consequences and anticipated disruptions.

4.10   From Theory to Practice

Naturally, there are no formal mechanisms in place to accomplish the sort of review and oversight mandated in human subjects research. Therein lies perhaps the greatest challenge to actually engaging the ethical dimensions of nanotechnology. The language of ethics is only as useful as the contexts in which it is used, the effectiveness with which it is employed, and the degree to which it is a valued dimension of societal considerations. Approaches most likely to be successful will have to accommodate the challenges of pace, complexity, and uncertainty with which we have been dealing throughout our discussion. Models for incorporating ethical considerations into key decision points will have to be able to respond quickly to an array of stakeholders and viewpoints so as not to be perceived as overly disruptive to the pace of innovation and time-sensitive demands of the marketplace. Likewise, successful models will have to be able to efficiently and effectively incorporate the challenges of complexity. This suggests the need for an array of commonly shared tools and resources that can detect patterns and make good use of prior insights and decisions so as not to reinvent the wheel with each new iteration of scientific discovery and technological advance. Finally, there needs to be some general consensus on the level of uncertainty we are willing and able to tolerate in terms of both risk and benefit. The current approach to societal impacts tends to roll nanotechnology into one, all-encompassing category. Rather than treating nanotechnology as a monolithic, all-or-none proposition, a proficient identification, sorting, and prioritization of issues can allow for faster progress in less problematic areas while encouraging restraint in those areas that present greater concerns or uncertainties.

In the next chapter, we turn to method and process in ethical analysis. The language of ethics is best employed within a systematic approach to problem-solving that identifies the known, poses the practical decisions that need to be made, frames the ethical considerations, and then uses ethics to facilitate the assessment and selection of ethically refined options for action.

4.11   Questions for Thought

1. Most of us have ways of thinking and responding to ethical dilemmas that align with the various concepts presented in this chapter. For example, my first instinct might be to do a cost benefit analysis (utility) or to simply focus on avoiding harm. I might be inclined to be loyal to my organization, but draw the line at lying or stretching

the truth. I might prefer fairness over any other duties. What is your ethical comfort zone and how does that impact your ability to identify ethical dilemmas and work through them with others?

2. Read several articles on nanotechnology and see if you can recognize the moral concepts being used in each piece, either subtly or overtly.

5

Method and Process in Ethics

5.1 Ethical Discernment and Dialogue

Whenever two good people argue over principles, they are both right.

—**Marie Ebner Von Eschenbach**

The term *discernment* can be defined as an "act or process of exhibiting keen insight and good judgment" (American Heritage Dictionary 2000).[1] It refers to a combination of skills or qualities, such as perceptiveness and acuity. Such qualities allow one to see between the lines, to recognize deep patterns, and to make mental connections between objects, relationships, and concepts. A discerning person often sees and understands what others do not. This ability of deep insight allows one to make better, wiser, more informed, and more prudent judgments about a topic or situation.

As I have argued in previous chapters, ethical analysis is best considered a method of critical thought. It can also be described as a process of discernment, requiring fine discrimination of subtle or even obscure concepts, and a consistency of thought that can be hard to achieve in the face of limited time, complexity, and inherent uncertainty. Furthermore, the application of ethical analysis goes well beyond the simple ability to reflect deeply on an issue. Ethical analysis enables basic decision making, problem solving, and conflict resolution. When done well, the process is characterized by a profound openness and sensitivity to multiple points of view, as well as a respectful appreciation for the complexities inherent in the dilemma at hand.

Whether applied individually or in a group setting, the ability to articulate the moral dimensions of an issue adds depth and substance to any discussion. With respect to nanotechnology, this enhanced depth and substance is particularly suited to helping address the complicating factors we have explored in earlier chapters. The pace of the external environment may be accelerating, but that does not alter the basic fact that decisions made too quickly are inevitably less than optimal decisions. Raising well-placed questions of utility, duty, justice, or intent creates much needed time and space for reflection and the testing of assumptions. At the same time, the more confident and consistent an individual or organization becomes in addressing these questions, the more quickly and efficiently moral dilemmas are resolved, and the less chance decisions will need to be revisited because they were incomplete or ill-conceived.

With respect to complexity, ethical dialogue positively invites it. All of the key abilities needed to cope with complex analysis and decision making are built into the process of ethical analysis, including the ability to think flexibly at a deep systems level. Those same questions of utility, duty, justice, and intent are avenues into the multifaceted relationships and outcomes that exist within complex adaptive systems. Perhaps the greatest challenge we face, when considering the societal impacts of nanotechnology and its converging partners, is the limitation of our ability to fully imagine what does not yet exist. However, ethics is, at its core, an exercise in imagination—a moral imagination that is rooted within the history of human experience, values, meanings, and insight.

Nanotechnology may pose new opportunities and new problems, but the underlying moral questions associated with radical or disruptive technological advancements will likely remain the same. This small element of predictability provides a consistent platform upon which we can begin to collaboratively build certain assumptions into our ongoing assessments. Nonetheless, the question of uncertainty still looms large. The higher the level of uncertainty in a situation, the less able we are to make decisions based on narrow assumptions of fact-based utility. Instead, we are challenged to project, as best we can, the outcomes of our decisions and the untoward consequences they may incur. The same assumptions that help us cope with complexity can guide how we think in the face of uncertainty.

In this chapter, we will focus on method and process. I will first outline some of the basic characteristics and abilities inherent in effective ethical analysis, followed by a general discussion of various approaches to process. One particular model will be outlined in depth and will serve as the basic outline for Parts Two and Three of the book.

5.2   Approaches to Ethical Discourse and Decision Making

As noted in the previous chapter, an ethical problem occurs in any situation in which there are moral factors to be considered in selecting an action. Furthermore, almost any situation has some moral element at stake. An ethical dilemma is a bit more involved since it entails a situation in which all choices have both beneficial and harmful, or desirable and undesirable effects. Where a true ethical dilemma exists, there are either compelling arguments both for taking an action and not taking that same action, or there are compelling arguments for taking two incompatible actions. Both situations invite ethical reflection, discernment, and discourse.

Ethical reflection is a process of weighing practical and moral insights. Ideally, ethical reflection is based on an informed understanding of the facts of a case or situation; the ability to recognize the practical problem(s) and decision(s) that need to be made; a similar ability to frame and evaluate the ethical questions from a variety of ethical perspectives; and the application of the resulting ethical insights to subsequent decisions and actions. Ethical reflection also includes the willingness to retrospectively evaluate the soundness of those same decisions and actions once implemented.

To be most effective, ethical reflection requires a number of important skills. Foremost is the ability to recognize personal biases, stereotypes, assumptions, values, purposes, and motives, as well as the effect they have on our ability to be truly objective about alternative points of view. The companion ability then is a certain independence and willingness to understand and consider alternative points of view fairly. **Intellectual humility** requires that we be able to admit that which we simply do not know, or about which we may have an incomplete or mistaken understanding. Openness to new information and unfamiliar points of view expands our moral capacity to imagine positions that may be ethically more sophisticated in their depth and breadth. Patience and persistence are two other requirements for sound ethical reflection. The more complex the problem or dilemma, the more tenacity one must have to sort out and prioritize the important elements, and then reconcile one's conclusions with prior positions and similar actions. Finally, moral reflection requires the courage to ask hard questions, challenge past assumptions, and respond to new insights with new positions and actions that may place us at odds with others.

The term **ethical discourse** is commonly used as a fancy term for conversation or discussion. Discourses can occur verbally or in written formats, but basically involve an exchange of thoughts and ideas. Interestingly, the Middle English equivalent, *"discours,"* refers to a process of reasoning. Thus, we can consider ethical discourse as a process of communication involving moral or ethical reasoning. For our purposes, it is ethical reflection done in, and for, the benefit of community. As with individual reflection, ethical discourse requires a similar set of abilities and attitudes. The first is the capacity to recognize the goals and interests of others and to be transparent about one's own goals and self-interest. Those familiar with the basic process of conflict resolution will quickly recognize these as critical steps that lay the basis for common ground. Likewise, the group's shared capacities for independence, fairness, intellectual humility, openness, moral imagination, tenacity, and courage are critical factors in the ability to conduct productive moral discourse.

Making difficult decisions at an organizational or policy level is a complex process that requires solid communication skills, a respectful attitude, and willingness on the part of participants to present and defend their individual positions. This process is both enhanced and complicated by the presence of diverse viewpoints and differing objectives. A variety of viewpoints help ensure that the group has considered the situation from more than one or two angles. On the other hand, strongly held competing positions can make building a consensus for action very difficult. The process can also be hampered by inexperienced participants, competing political agendas, an absence of leadership, or a lack of clear direction. Finally, rapid pace, complexity, and uncertainty present powerful pressures to move ahead before a well-reasoned consensus is reached. As with any other form of conflict resolution, it is perhaps the nature of ethical discourse that, in reality, it will fall below the ideals described here. However, this reality does not diminish the value of cultivating the ideal in order to build the capacity for ethical discourse as we move into the future. Emerging technologies are among the most pressing issues we face, now and well into the future, suggesting that our capacity to effectively assess them must also continue to emerge and evolve.

Such capacity for ethical discourse in a group requires at least three specific conditions. First, the group must be willing and able to explore the ethical implications of a situation under consideration—in other words to ask the hard questions. There is a natural tendency to focus narrowly on the practical problem(s), potential solutions, and specific decisions to be made. To consider the ethical dimensions of a situation requires the additional steps of generating ethical concerns, framing ethical questions, and conducting ethical analysis. This extended process requires a commitment of both time and energy by members of the group.

A second condition is that individuals within the group must be able to generate ethically grounded questions and positions. This task requires some familiarity with the language and conceptual content of ethics; skill in identifying concerns from differing ethical viewpoints; and then crafting each concern in the form of a question. Once the questions have been crafted, the work of answering them in a way that provides genuine guidance to the decision making process begins.

Therefore, the final condition involves the group's ability to explore diverse positions and achieve consensus on decisions in which ethical considerations are clearly addressed. This aspect requires a respectful atmosphere in which opposing viewpoints are welcomed and participants feel secure in expressing their opinions openly. In addition, the group must be able to identify areas of common ground in order to create a basis for consensus. All in all, these conditions are not easy to achieve in the presence of competing agendas, pressing time frames, and either strongly divergent or overly homogenous viewpoints.

The combination of experience and ethical reflection plays an important role in moral analysis and decision making. However, when an individual or group is confronted with a moral problem or dilemma, particularly a situation that has not previously been faced, it is also helpful to apply a systematic approach to consideration of the problem. Even the most intuitive decisions benefit from a comprehensive and consistent process of analysis and decision making. An ethical decision model is a step-by-step framework for making decisions and there are a variety of such models for problem solving and decision making (Johnson 2005).[2] The specific model you select for use is far less important than the fact that you have a framework and process that cultivates rigor, efficiency, and consistency. Each model tends to use ethical theory and concepts a bit differently and, just as some ethical theories tend to be a more comfortable fit for individuals and organizations, some models will make more sense than others.

Minimally, an acceptable model for ethical reflection and decision making will accomplish the following objectives.

- Gather the necessary information to understand the situation as fully as possible.
- Identify the practical problems and decisions that need to be made in order to take action.
- Identify the ethical issues and frame the ethical questions that are raised by the issues.
- Construct the ethical arguments and counterarguments that answer the ethical questions posed, and arrive at a set of ethical criteria for evaluating specific options for action.
- Generate a range of options for action and test each option using the ethical criteria developed in the analysis.
- Address the practical problem by selecting the option for action that best addresses the combined practical and ethical issues raised.
- Evaluate the outcomes of the decision and adjust future responses accordingly.

Obviously, the *more* complete our understanding of the issue, the *more* stakeholders at the table, the *more* ethical dimensions explored, and the *more* creative we can be with our options for action, the greater the likelihood that our actions will effectively mediate both the practical and ethical concerns. Unfortunately, each of these *"more"* statements also increases the practical challenges entailed in conducting a comprehensive assessment. However, what is most important to take away from this list of objectives is that it does not in any way distance itself from the pragmatic considerations raised by the issue at hand. Gathering information, identifying problems, generating options, and taking action are all going to happen anyway, so the integration of ethical analysis is not an unreasonable intrusion on everyday decision making. As argued previously, we already do ethical analysis, although per-haps without explicitly recognizing it as such. Nonetheless, ethical dimen-sions are already implicitly embedded within practical decision making and simply making them explicit is a relatively modest goal, at least in theory. For now, let's turn to a more detailed discussion of what might be included in a model of ethical analysis.

5.3   A Model for Ethical Analysis

For our purposes in the applied sections of the book, I will follow a basic model of ethical analysis based on the following elements.

- Describe the Context
- Clarify the Purpose
- Frame the Ethical Questions
- Point/Counterpoint
- Assess Options for Action
- Find Common Ground

Notice that I have not described or numbered these elements as "steps." In fact, I hesitate to even use the term steps since it implies an overtly linear process or progression. Ethical reflection is, at its best, more iterative than strictly linear in nature. Each element can readily lead to a reconsideration of the prior elements. New information and insights should be readily absorbed back into the process, circling as needed to obtain consistency and consensus. There is a definite tendency for any particular ethical question to raise several more questions, leading my students to frequently complain that their heads hurt by the end of class. And, this tendency to spin one's wheels among competing ethical demands is a perfectly valid concern expressed by stakeholders who fear becoming mired down in endless rhetoric and circular arguments. On the other hand, when common ground can be found, a firm direction can be established for moving forward, even if not quite all issues have been resolved to everyone's satisfaction. What this achieves is a justification based on more than narrowly utilitarian or self-interested reasoning. This justification can serve to increase the level of trust and buy-in for potential opportunities, as well as increased tolerance for those potential risks and harms that have been openly discussed and evaluated.

Of course, nothing is ever simple. Nanotechnology and its technological counterparts cannot be reduced to a one-shot analysis. There is no one context that encompasses nanotechnology. The facts are dynamic as are the various settings in which nanotechnology is emerging. Nanotechnology also cannot be described as having a single, all-encompassing purpose. In fact, efforts to broadly portray the goals of nanotechnology as simply another leap in technological progress that is intended to improve the quality of our lives, are misleading and unhelpful. Different applications of nanotechnology raise different societal implications, and are accompanied by different ethical questions and different ethical arguments. Likewise, each application or problem under consideration, within its particular setting, may generate unique and different options for action. The suggested options that we choose between totally pulling the plug, slowing everything down with regulation, or barreling full steam ahead are likewise unhelpful. Each decision made will be, at one level, uniquely suited to its own situation and evaluated on its effectiveness in that specific context. However, at the same time, effective decision making that includes explicit ethical analysis and justification can set powerful precedents. In turn, such precedents can be incorporated to streamline subsequent decisions.

As suggested in the prior section, the model presented here is based on an extension of the standard problem-solving model of management. It includes the basic elements of defining the problem, gathering information, generating options, and selecting an action. It can be applied to a specific case or a broader issue. Let's turn to a more detailed examination of process and content within each element.

5.4 Describing the Context

In order to describe the context in which an issue either is or should be considered, one must gather relevant information consisting of the factual elements of the case or topic. The term "factual" is used broadly here to include both objective and subjective information. For example, when considering a case in bioethics, a patient's preference to refuse treatment may be based entirely on subjective interpretations and beliefs of the patient, but the refusal itself and the reasons for it must be treated as an objective fact. With respect to nanotechnology, there is so much we don't know that we are forced to operate on the basis of theoretical possibilities, statistical probabilities, and informed predictions for everything from the science itself, to the chances that any specific application will materialize, to the trajectory of the market, and so forth. Under such circumstances, it is not possible to know everything we might like to know in order to fully assess the implications or impact of an issue. In fact, we can count on rapid pace with its subsequent short time frames, increased complexity, and high levels of uncertainty to be contextual givens in any situation we might consider.

Nonetheless, there is much we can gather and, over time, a concerted effort to document societal implications discourses and outcomes can generate a body of knowledge and assumptions that we can draw upon for subsequent considerations (Bennett-Woods 2007).[3] Obviously, not all factors will be as relevant to all applications and issues. In many instances, the range of factors may be quite limited. In other instances, most or all factors will hold some relevance to the analysis. For illustration, let's briefly explore the potential range of information we might want to consider in a comprehensive deliberation regarding a particular issue related to nanotechnology.

5.4.1 The Scientific and Engineering Context

When assessing some aspect of nanoscience or nanotechnology, relevant facts should obviously include objective descriptions of the science itself, research findings, and proposed applications. As noted above, such a description might also include assumptions about specific outcomes based on common experience, statistical likelihood, or informed predictions. In addition, what is not yet known or fully tested may be equally as important as what is known.

This is an opportune moment to insert a comment regarding how the science is understood and communicated. A common criticism from scientists and others regarding the portrayal of societal implications is that both critics and supporters misrepresent the science. This tends to lead to either overly optimistic or highly pessimistic perceptions of nanotechnology. The ongoing debate about molecular manufacturing is a good example of both informed disagreement about what is actually possible, what is possible but unlikely, and what amounts to largely sensationalized overstatement from misinformed commentators (Berube 2006).[4] If nanotechnology is to be subject to a broader societal consideration, it would be incumbent upon the scientific community to provide accurate, understandable lay descriptions and explanations of the scientific or technological activities and goals being considered. It could be noted here that one of the by-products of the proposed model of discourse is the much-needed education of policy makers, regulators, interest groups, the general public, and other stakeholders.

At minimum, the following would be of interest to any general consideration of a substantial advance in nanoscience or nanotechnology.

- Lay descriptions of the basic science, research findings, and potential applications
- Specific goals and intended outcomes of conducting the science or developing proposed applications
- Potential convergence with or enabling effects in other technologies
- Projected time lines
- Safety concerns—both short- and long-term
- Technical barriers and unknowns
- Potential misuse of the applications

5.4.2   The Legal, Regulatory, and Policy Context

The legal, regulatory, and policy context provides insight into the current legal and regulatory status of this and similar technologies, as well as legal and regulatory precedents that may be of some use as analogies. An understanding of current regulatory requirements, or the absence of regulation, is of particular importance to investors and the business community. Current or likely political considerations and policy issues should be identified, and may overlap with public perceptions. The following factors are examples of information that might be useful.

- Current regulatory restrictions and guidelines, including any under consideration
- Relevant legal issues related to patent law, liability, technology transfer, etc.
- National and international policy considerations
- International regulatory activities, concerns, or pressures
- Current and prior efforts at self-regulation
- Industry influence
- Competing priorities for funding and support
- Regulatory and policy precedents

5.4.3 The Economic and Market Context

An assessment of the potential market for a new technology helps to establish its likely trajectory and identify key stakeholders. Nanotechnology is routinely billed as an economic engine of unprecedented magnitude (Roco and Bainbridge 2001).[5] Here again, there is much speculation that projections are overstated and sensationalized (Berube 2006).[4] In either case, it will be very easy for economic and market considerations to dominate the societal implications dialogue, so reasonably accurate projections and a certain transparency will be important to this category.

- Investment and revenue projections
- Timelines
- Industries affected and primary stakeholders
- Competitive pressures
- Consumer interest and support
- Consumer access
- Consumer concerns and product liability issues
- Workforce issues
- Projected effects on existing markets
- Global market considerations

5.4.4 The Environmental, Health, and Safety Context

Although this category overlaps somewhat with several other categories, it is increasingly becoming a strong enough focus to warrant its own designation. As noted previously, the environmental, health, and safety (EHS) category of the NNI is commanding a growing share of the societal implications research budget as well as supplemental, discretionary funding from certain agencies. Information that might be of importance here includes:

- Known or suspected hazards
- Current risk assessment activities and findings
- Safety measures in effect

5.4.5   The Public Context

This final catch-all category is the repository of all the remaining miscellaneous factors that may need to be considered in an informed assessment. Relevant factors here will vary widely based on the particular issue under consideration and the setting. Examples of factors in this category might include:

- Public understanding and perceptions
- Media portrayals
- Activities of non-governmental organizations, consumer interest groups, and grassroots activism
- Religious and cultural issues
- Significant external events (such as another major terrorist attack or a severe disruption of stock markets)

Ethical analysis that is applied narrowly to issues in only one or two of these categories will fall short. The intersections between categories overlap in dynamic and often unpredictable ways. Rapid pace assures us that the "facts" will be a moving target as the science advances and technology continues to emerge. In much the same way a business performs ongoing market scans, truly dynamic technology assessment will require an enhanced capacity to detect important shifts in each of these categories on an ongoing basis. Coupled with this ongoing monitoring is the ability to recognize opportunities for revisiting the most recent analysis in light of new information.

5.5   Clarifying the Purpose

Clarifying the purpose of the ethical analysis is the process of simply providing a statement of the practical problem or problems that need to be addressed. Problems often take the form of decisions that must be made in order to take action. The focus of this particular element of the model is to simply state the decision(s) that needs to be made rather than explore what specific action should be taken or the justification for any one particular course of action. In fact, if the problem is stated too narrowly, as in terms of one solution or as an all-or-nothing proposition, then the problem statement itself tends to bias the analysis and limit the brainstorming of options later in the process.

There are several naturally occurring decision points inherent to the process of science and technology (Bennett-Woods and Fisher 2004).[6] These decision points provide good examples of straightforward problem statements. For example, consider a researcher's request for funding. The practical problem is simply, "Should we provide initial or continued funding, and at what level?" Advancements in science and technology require the sustained investment of public and/or private funding. Initial and continued funding is routinely evaluated, at multiple points, and granted or denied based on technical, economic, and other practical considerations. In addition, one can also recognize the potential for ethical considerations. At minimum, the funding source will likely weigh basic utilitarian concerns of balancing the cost/investment with the likely benefits, particularly when considering competing funding proposals. In the face of scarce resources, there is an unavoidable harm caused when determining to fund one promising line of research over another, insofar as those who would have benefited from the research not selected for funding will not have that opportunity. In addition, the merits of the various proposals may involve having to consider the most pressing issues. The funding priorities for medical research continually reflect determinations about how best to allocate scarce funds between research into a chronic disease that affects many people or a deadly disease that affects far fewer; whether to fund research that leads to prevention or research focused on cures; or whether to fund research into costly new medical technologies at all, given the cost constraints of the current health care system. Another potential ethical consideration is fidelity to the extent that funding decisions may reflect certain loyalties to individuals, the community, or other stakeholders.

If our goal is to integrate a broader range of societal issues into the fabric of the research process itself, and in a manner that is both practical and useful to the researchers and decision makers themselves, then the funding proposal is as good a place as any to start. By adequately addressing ethical questions at the initial point of funding, subsequent decision points are less likely to surface questions that have not already been considered at some level.

Another common decision point in the science and technology continuum poses the question, "Should we allow the transfer of knowledge and, if so, to whom?" Techno-scientific knowledge is generated within a complex and largely unpredictable system of relationships. Every transfer of knowledge has the potential to spin off in unanticipated directions. This potential returns us to our earlier discussion of the role of scientists and engineers as moral agents and to what extent they are, as the originators of knowledge or a particular technological application, morally responsible for its subsequent use. A decision to transfer knowledge should perhaps be made with a willingness to remain morally committed at some reasonable level for potential applications.

A final example of a common decision point in science and technology considerations can be described in the question, "Should we regulate and, if so, to what extent?" This question primarily addresses potential or known outcomes that are harmful or otherwise undesirable. For this purpose, regulation can be defined broadly as any limitation imposed on the process, either self-imposed by individual researchers and corporate entities, or externally imposed by governmental oversight. On the other hand, the regulatory process can also be used to encourage or reward a particularly promising line of research and development by providing incentives or lifting existing regulatory obstacles.

The problem statement establishes the scope of the decision or decisions to be made. By being clear on the scope, and by crafting the problem carefully, one can more easily retain the proper focus while performing an analysis or assessment.

5.6 Framing the Ethical Questions

At this point, we are ready to identify the most significant ethical issues by stating them in the form of one or more ethical questions. The goal here is to formulate the basis of analysis and justification for taking or not taking a particular action. The best way to do this is to brainstorm a list of issues and questions using the assumptions provided by the various ethical principles and theoretical approaches. In the course of this process, we are likely to identify a variety of interesting, but quite different ethical issues and questions. Some questions can be determined to be foundational to other questions. Some questions may be of only secondary interest. It is helpful to identify the one or two ethical questions believed to be central to the issues at hand. In fact, this is a first step in developing consensus for a solution, since obtaining agreement on core questions will focus subsequent discussion on issues that everyone sees as important. Secondary questions can be retained and used to stimulate both supporting arguments and counterarguments.

Well-formulated ethical questions directly suggest appropriate principles and perspectives. For example, a question involving an inherent obligation can imply a duty-based or rights-based argument supported by one or more specific principles. A question regarding the greater good or cost/benefit considerations will likely imply a utilitarian argument. While your primary arguments are constructed in response to your primary questions, secondary questions may point to other principles and theories that can be considered in support of either arguments or counterarguments.

For illustration, let's consider questions that might be raised with respect to an advance in nanotechnology that enables precision genetic repair and manipulation of human DNA. An interesting range of ethical questions can be quickly generated to include:

- Will the likely or potential benefits of genetic repair and manipulation outweigh the likely or potential harms?

- Is the cost of development proportionate to the number of people who are likely to benefit?
- Do we have a duty of fidelity to current and future patients to develop genetic therapies as quickly and completely as possible?
- Is our intention limited to the repair of known genetic defects or do we also intend to manipulate DNA for purposes of genetic selection or enhancement? Is there a morally relevant difference?
- Will limited availability of genetic repair and manipulation, due to cost and access, increase existing social barriers and injustices?

5.7 Point/Counterpoint

The ethical principles and/or ethical frameworks implied by the ethical questions posed now become the basis for conducting a focused ethical analysis or constructing an ethical argument. An **argument** can be generally defined as a course of reasoning intended to prove the rightness or wrongness of a particular action or position. By extension, the **counterargument** is an alternate course of reasoning that challenges the logic, consistency, and/or basic assumptions of a particular argument. Using your ethical question(s) as the basis, you can formulate and test possible answers with reference to the various principles and theoretical perspectives. The initial focus should be on the primary question(s) and then on any secondary questions that might further enhance or refine the position.

The counterargument is an analysis that directly challenges your initial argument. The ability to convincingly articulate counterarguments is extremely valuable. First, it allows one to anticipate and understand the views of those in an opposing position. It also forces the consideration of exceptions to a position. It would not be a dilemma if there were not compelling arguments on both sides, which suggests the possibility that the strength, tone, and direction of certain arguments may need to be modified to ensure logical consistency or to address reasonable exceptions. Finally, well-constructed counterarguments can lead to the generation of additional options for action that might not have otherwise been considered.

5.8 Assessing Options for Action

This element is quite deliberately placed near the end of the decision model. Our natural tendency as problem solvers is to jump to possible solutions early in the process. However, this early focus on solutions can artificially bias and limit both the ethical arguments and the options for action that are generated and considered. For example, identifying a preferred option for action early in the process can subvert the analysis into a narrowly targeted justification for a convenient, comfortable, familiar, or self-serving approach, rather than an open-ended search for the optimal action. Therefore, the goal here is to brainstorm a wide range of action steps and then test each action against the arguments and counterarguments. It is important to remember that taking no action is also an option that has its own set of moral implications.

5.9   Finding Common Ground

Ultimately, the goal is to reach a point at which an informed and well-considered option for action or combination of action steps can be selected and implemented. However, the model does not stop there. In order to complete the process, one must return to reflect on the outcome of your deliberations and action to consider whether the action remains justified once it has been implemented. Reflection might include the following questions:

- Did I/we identify and consider all relevant information?
- Did I/we correctly identify and state the practical problems and decisions to be made?
- Did I/we correctly identify and frame the primary ethical issue(s) and question(s)?
- Were there important ethical principles and perspectives I/we failed to consider?
- Were there viable options for action that I/we did not consider?
- Would I/we make the same decision again?
- How might I/we approach a similar situation in the future?

This completes a very brief overview of one approach to method and process in ethics. Failure to give adequate attention to any one element of the model can limit the quality and integrity of the final decision. On the other hand, something as simple as posing one solitary ethical question can positively alter the depth and direction of a decision-making process, so slavishly adhering to any one model is not necessary in order to garner some benefit from integrating the language and process of ethics into everyday activities. Initially, the model may appear lengthy and cumbersome. Experience and a growing knowledge base do allow increasing efficiency over time. The discipline of routinely considering ethical questions and perspectives in daily decision making eventually enables faster construction of sophisticated justifications that tend also to become more refined and consistent over time. This practiced effectiveness is the nature of any form of critical thought and reasoning. The skills of ethical analysis carry over into all forms of problem solving.

5.10   Pragmatic Considerations

The creation of workable opportunities for ethical discourse on nanotechnology is a major challenge. The proliferation of guidelines and principles, analytic frameworks, and models for dialogue is necessary for ethical assessment, but will be of limited practical use unless they can be effectively and efficiently incorporated into the existing systems and culture of the science and technology enterprise. In an ideal world, technology assessment would be accomplished in a balanced, collaborative, and dynamic consideration of opportunities and concerns. It would draw efficiently and effectively from diverse knowledge bases and in a time-sensitive manner so as not to become an undue burden to the forward progress of science. Developing such an approach to nanotechnology will not be easy, but the opportunity to create practical and effective strategies for coping with the challenges presented by nanotechnology will benefit all future considerations of emergent technologies (Bennett-Woods 2007).[3]

Efforts are being made to create such opportunities. For example, Fisher and Mahajan (2006)[7] studied the feasibility of introducing a decision protocol that integrates societal considerations into nanotechnology research decisions in the university setting. In this extremely small study, the researcher functioned as an "embedded humanist," a member of the research group with expertise in both humanities and policy research. Sample potential concerns regarding nanotechnology were introduced to the designated participant via semi-structured interviews during scheduled meetings to document the progress of graduate research projects. The researchers concluded that the decision protocol did stimulate awareness on the part of the graduate students and resulted in altering certain decisions in response to societal concerns. In essence, the simple act of posing questions enhanced the scientists' ability and inclination to reflect more broadly on the aspects of their work that raise potential societal concerns. The process was not characterized as time-consuming or burdensome. It did not appear to impede the normal flow of the research process. It is entirely consistent with the method and process proposed in this chapter.

Of course this very limited example is a far cry from broadly incorporating formal ethical analysis into the decision-making processes of research teams, funding sources, corporations, policy makers, and regulatory agencies, let alone general public discourse. The practical barriers are many, beginning with the tendency to regulate technologies either at the beginning or ending stages of their development cycles, while largely ignoring the rest of the development process itself. Regulation early in the development cycle often takes the form of "yes or no" funding decisions and priorities that may or may not be supported by a sound ethical justification. Regulation that occurs at the end of the development process, especially once a product has hit the market, is often too late to be fully effective (Bennett-Woods and Fisher 2004).[6] Idhe (1993) makes this point, stating that applied ethics tends to come *"too late....after all the technologies are in place."*[8] As a result, there are calls from nanotechnology proponents to move ethical analysis "upstream" and much earlier in the science and technology development process (Roco and Bainbridge 2001; Royal Society 2004).[5,9] For example, the Royal Society and the Royal Academy of Engineering makes this early consideration of ethical and societal implications a priority embedded within a number of formal recommendations in their 2004 report entitled *Nanoscience and Nanotechnologies: Opportunities and Uncertainties.*

> The upstream nature of most nanotechnologies means that there is an opportunity to generate a constructive and proactive debate about the future of the technology now, before deeply entrenched or polarized positions appear. (p. 82)[9]

Related recommendations include interdisciplinary research into the social and ethical issues, required incorporation of the consideration of social and ethical implications of advanced technologies into the formal training of all research students and staff, and adequately funded public dialogue. Their final recommendation reiterates the need for early assessment.

> We recommend that the Chief Scientific Advisor should establish a group that brings together representatives of a wide range of stakeholders to look at new and emerging technologies and identify at the earliest possible stage areas where potential health, safety, environmental, social, ethical, and regulatory issues may arise and advise on how these might be addressed. (p. 87)[9]

Tepper (1996)[10] points out that it makes sense to do basic assessment early in the research and development process when it is easier and more cost effective if a course change is warranted. Likewise, it makes good sense to integrate ethical assessment into the natural flow of the development cycle (Bennett-Woods and Fisher 2004).[6]

Although this all sounds straightforward, the development cycle of a technology is relatively complex (Pielke and Byerly 1998).[11] It is not a simple linear process. Furthermore, nanotechnology provides the additional challenges of being both interdisciplinary and aimed mostly at enabling other technologies with which it converges (Bennett-Woods and Fisher 2004).[6]

Let's expand a bit on the non-linear nature of the research and development process. The classic model of techno-science development, as articulated by Bush (1948),[12] is a linear progression in which knowledge generated in basic research is then used in applied research to enable technology development and result finally in some societal benefit(s). According to the linear model, knowledge and its products generally "flow" from each stage to the next more or less automatically. The model suggests a clear separation between basic and applied research with the stream of knowledge moving more or less in one direction from basic to applied. Furthermore, basic research tends not to be constrained by practical or other external considerations, while applied research is somewhat more limited due to the expectation it will lead to practical applications (Branscomb and Florida 1998).[13]

While the idealized linear model can provide insight into the overall process of science and technology, it is also open to a number of objections. For instance, knowledge does not necessarily "flow" in the direction dictated by the model. New information gained at any point in the R&D continuum can feed back into earlier phases, forcing new insights and directions. The idealized form of the model tends to discourage critical assessment by separating the science from its likely or potential technological outcomes as was discussed in earlier chapters. The observation that knowledge does not flow smoothly in a single direction, as suggested by the model, creates an additional barrier to timely and complete science and technology assessment.

However, there are key decision points in the continuum of science and technology research and development at which assessment makes sense and could reasonably include broad ethical and societal considerations (Bennett-Woods and Fisher 2004; Bennett-Woods 2007).[6,3] For example, at the beginning of the cycle and various points throughout the continuum, decisions are made regarding whether or not to fund new or continued research and development. With increasingly constrained budgets for research and development, we may be obligated to prioritize some research initiatives over others on the basis of a greater good. Another key decision point that can arise throughout the process is whether or not we should transfer knowledge developed for one purpose and in one industry sector to another industry and possibly for an alternative purpose. A common example is the transfer of technologies originally developed for military purposes into the private sector. The navigation system in your new car has its origins in military applications of global positioning technology. Nano-enabled technologies developed for military surveillance are likely to make their way into civilian use as well. This raises legitimate questions, such as control, in terms of questionable applications, and fairness regarding who should benefit from research conducted at the public expense. A final decision point that should also come up throughout the development process involves whether or not we should regulate either the science itself, the transfer of knowledge, or the technological applications developed. This question is often posed only after serious environmental, health, and safety effects have been observed in a product or process already on the market (Harremoes et al. 2001),[14] undermining the value of regulatory assessment and oversight in promoting consumer safety and informed consent in the face of corporate interests.

These decision points represent junctures within the development process that are also logical targets for ethical consideration and effective intervention, but such intervention requires a technology development process model that is more sensitive to the potential intended and untoward research outcomes. Identifying likely decision points requires discarding the concept of "pure science" in favor of science tied inevitably to its potential applications and accompanying societal outcomes. Critical decision points would then involve any point during the research and development process that could influence decisions made with regard to potential "downstream" ethical and social outcomes. Assessment at such points would include integration of new scientific and technological information along with values and other societal considerations.

Under some circumstances, the flow between basic science and technology development can be relatively seamless and remain under the control of a single lab or project. However, the interdisciplinary nature of nanotechnology, along with the factors of complexity and pace, make it more likely that the process will be relatively discontinuous, such as when results are applied or modified by different researchers at different times, in different projects or institutions, and sometimes without the knowledge of those initially responsible for an earlier phase. The increased pace of discovery and development also makes it more likely that phases of the model will overlap, occur simultaneously, flow backwards, or be delayed. A development model that effectively incorporates broad societal assessment must identify likely decision points based on pragmatic considerations of both process and outcomes. It must be flexible and practical, as well as consistent with the natural flow and relative unpredictability of the full continuum of research and development (Bennett-Woods and Fisher 2004).[6]

In theory, it may be easier to make changes early in the research process; however, in reality, it may also be much harder to identify exactly what changes need to be made at these stages as opposed to during the applied stages when outcomes are better defined and more tangible. Therefore, although there are increasing calls for early intervention, there is no recognized approach to altering existing R&D practices accordingly. In the proposed model, intervention would be guided by specific "outcomes assessment and intervention points" (OAIPs).

With potential OAIPs identified, dialogue regarding ethical considerations can be seamlessly introduced into practical decision making regarding funding, regulatory implications, and potential limitations on further development or transfer that routinely (though often implicitly) occurs at various points along the R&D continuum. However, incorporating ethical and societal considerations and analysis at such points also requires a broadening of the process to include stakeholders not normally involved at such decision points. This expansion of the decision process can be accomplished in various ways, such as by establishing internal or external review boards (similar to internal review boards, or IRBs, in human subjects research), incorporating social scientists into project teams, and use of external consultants to facilitate dialogue.

Whatever methods are used to incorporate ethical analysis and societal considerations into the science and technology research and development continuum, change will require commitment on the part of funding agencies, scientists and engineers, corporate entities, and regulators to actively support and participate in the process. In fact, re-envisioning the R&D process to broadly include more direct responsibility and accountability for its outcomes represents a shift in culture within the scientific enterprise that is likely to take time and much concerted effort. On the other hand, simply introducing the basic language of ethics and elements of ethical process and method into the daily activities of scientists, engineers, and their sponsors can work relatively quickly to build a strong foundation for a more thorough and integrated model in the future.

5.11 Questions for Thought

1. Select a controversial application of nanotechnology and apply the basic decision model presented in this chapter.

2. Raise the issue above over lunch with a colleague or group of colleagues and try to facilitate a similar process during the discussion. Observe the various points of view and rationales. Practice finding common ground from which options for action can be implemented.

PART TWO

Emerging Issues

The future enters into us, in order to transform us, long before it happens.

—Rainer Maria Rilke, German poet (1875-1926)

6

Nanomaterials and Manufacturing

6.1 The Vision of Nanotechnology

> The emerging fields of nanoscience and nanoengineering are leading to unprecedented understanding and control over the fundamental building blocks of all physical things. This is likely to change the way almost everything—from vaccines, to computers, to automobile tires, to objects not yet imagined—is designed and made.
>
> **—Nanotechnology: Shaping the World Atom by Atom[1]**

The vision of nanotechnology is at once simple yet sweeping in its impact. The quotation above is the opening statement of a brochure published by the U.S. National Science and Technology Council in 1999. This brochure, among other publications, signaled the start of a major initiative by the United States to shepherd the technological promise of nanotechnology into existence. Similar documents have emerged across the globe in at least 30 other countries, as governments and their scientists mobilize in the race to reap the benefits of what promises to be the next great technological revolution.

A widely cited model proposed by Mihail Roco (2004),[2] Senior Advisor for Nanotechnology at the National Science Foundation, depicts a timeline for the emergence of the first commercial prototypes of nanotechnology across four generations. The first generation, beginning in 2001, involves what he terms "passive nanostructures" that are used to modify stable macroscale properties and functions. Examples include nanostructured coatings, dispersion of nanoparticles, and bulk materials, such as nanostructured metals, polymers, and ceramics. The second generation began in 2005 and moved toward the use of "active nanostructures" for mechanical, electronic, magnetic, photonic, biological, and other effects. These structures are generally integrated into microscale devices and systems, and include new transistors, nanoelectronic components, amplifiers, targeted drugs and chemicals, actuators, artificial muscles, and adaptive structures. The third generation of products is estimated to begin arriving in 2010. It is described as "systems of nanosystems" that make use of syntheses and assembling techniques, such as bio-assembling, robotics with emerging behavior, and evolutionary approaches. By overcoming challenges related to networking and architecture, these applications will include directed multiscale self-assembly, artificial tissues and sensor systems, quantum interactions within nanoscale systems, photonic processing of information, and early platforms for converging technologies involving nanotechnology, biotechnology, information technology, and cognitive science (NBIC). Finally, between 2015 and 2020, it is estimated that we will see the emergence of "molecular nanosystems" in which each individual molecule is engineered as a device with a specific structure and function. This generation will enable nanoscale machines, quantum control, biological nanosystems, and human/machine interfaces.[2]

Each succeeding generation represents a greater level of control, manipulation, and effect at the nanoscale. As such, each generation represents a technological leap that moves us further from our current baseline of experience. However, here is where we begin to see a fragmenting of the vision as alternative visions begin to emerge. The infamous article entitled *Why the Future Doesn't Need Us*, published by Bill Joy in 2000, set off alarm bells regarding work at the nanoscale as well as other 21st century technologies.[3] The result is that where one person sees personalized medicine, another sees the emergence of human beings enhanced to the point they no longer consider themselves of the same species. Where one person sees sustainable manufacturing, another sees toxic degradation of the environment and the permanent loss of manufacturing jobs. Where one person sees military supremacy and peace, another sees unprecedented weapons of mass destruction and a police state in which personal privacy no longer exists.

The initial vision is further clouded by lack of common technical definitions, market hyperbole, political agendas, the speculative nature of nanotechnology, and a general lack of public engagement regarding science. It is here we begin to explore a few of the ethical implications of nanoscience and nanotechnology, as they are likely to be applied in the real world.

6.2   The Context Described: Scenarios
in the Nanotech Marketplace

For the reasons discussed in the prior section, combined with the rapid pace, complexity, and uncertainty discussed in previous chapters, there is no clear basis on which to even hazard a guess as to how nanotechnology is actually likely to unfold in the marketplace. Forecasts of the size of the nanotechnology market vary wildly from a few billion to nearly $3 trillion by 2015.[4] Investment in nanotechnology research and development is a bit easier to gauge. Worldwide nanotechnology funding is projected to be over $27 billion between 2006 and 2010.[5] Japan leads in spending, followed closely by the United States; however, the combined European countries will exceed both Japan and the United States.[5] Japan and the United States also lead in patent applications.[5]

COMMENTARY

Rocky Rawstern

Fire, electricity, silicon; the bases for technologies that have had a profound impact on society. Nanotechnologies will have an even greater impact. Like virtually all technologies, nanotechnologies present a double-edged sword; they are neither good nor bad, and can be put to uses both positive and negative. How we prepare for nanotechnologies will determine the extent to which we are able to maximize their best potential and minimize the worst.

All stakeholders (that's you, me, and everybody) need to have at least a basic understanding of nanotechnologies and participate in the decisions leading up to their use. The 21st Century will be a time of unprecedented change; change enabled by our understanding of the nanoscale. The decisions we make today will determine the future for generations to come.

No informed person doubts that developments at the nanoscale will be significant. We debate the time frame, the magnitude, and the possibilities, but not the likelihood for large-scale change. The least-speculative views suggest that we're in for changes of an order that justifies—if not demands—our undivided attention. Will we be ready?

Rocky Rawstern, Founder, Access Team. Rocky served as Editor of *Nanotechnology Now* (NN) for more than five years, and was instrumental in the creation of, and updates to, the NN website, products, and services. During that time he collected, categorized, and posted over 20,000 nanotechnology and nanotech-related news articles and press releases, and interviewed hundreds of stakeholders. For "outstanding journalistic or other communication endeavors that lead to a better public understanding of molecular nanotechnology and its high social and environmental impact," Rocky was awarded the 2005 Foresight Prize in Communication.

There are already a number of consumer products on the market that claim to use nanotechnology, although not all products using the label "nano" actually exploit nanoscale properties. Most products currently on the market employ interface effects such as scratch resistance, water repellency, flame retardancy, and UV protection. A few products incorporate quantum mechanical effects or complexity.[6] Examples of products include stain and odor-resistant and UV protected apparel, personal care products, such as sunscreens and toothpastes, consumer electronics, sports equipment, and home improvement and household products.[6]

The scope of economic sectors generally predicted to be affected by nanotechnology is quite broad. The impact on industrial manufacturing, materials, and products will be far-reaching since it will fundamentally change the way materials and devices are produced.[7] Other economic sectors commonly mentioned include medicine, transportation, energy, agriculture, electronics, chemicals, pharmaceuticals, national security, and space exploration.[7] Sustainability, with respect to environmental concerns, is also generally included in the list of economic targets, in contrast to concerns regarding nanotechnology as an environmental hazard in and of itself.[7]

The societal impact of nanotechnology goes beyond new economic opportunities in the various sectors. It also holds great implications for all aspects of the economy including the relative price of products, the initial cost of investment, global competitiveness, wages, and the potential loss of existing jobs. Other issues that are frequently raised include workforce education and training; workplace safety; legal challenges related to intellectual property and patenting; and public understanding and acceptance.[7] A final issue to mention might be the NNI itself and the unprecedented collaboration between government, business and industry, and academia that raises numerous questions of conflict of interest and the stewardship of public dollars.[8]

COMMENTARY

Meyya Meyyappan

Nanotechnology is not about simple extrapolation of length scale from micro to nanodimensions. It is more about taking advantage of novel properties that arise solely due to the nanoscale and producing useful or functional components. Physical, chemical, mechanical, electrical, magnetic, optical, and other properties change when materials reach nanodimensions. Since all existing products are based on one or more of the properties of the material used to construct that product, we can then expect the impact of nanotechnology to be on all the existing products or processes we have at present. By considering economic sectors, the impact of nanotechnology will be felt on computing, memory, data storage, communications, electronics, photonics, materials and manufacturing, chemicals and plastics, energy, environment, transportation, health and medicine, national security, space exploration, etc. Nanotechnology is, therefore, not any single technology and we should speak in terms of nanotechnologies. However, it is more appropriate to recognize it as an **enabling technology**. Other examples of enabling technologies from the past include textiles, railroad, automobile, aviation, and computers.

Nanotechnology, as an enabling technology, is expected to impact all sectors of the economy. The timeline for this impact depends on the degree of technical challenge, as well as the needs and common practices of each sector. For example, application of nanomaterials in consumer goods, such as tennis rackets and cosmetics, has already happened in the marketplace.

On the other hand, replacing silicon in computer logic is not only technically challenging, but the enormous investments in silicon technology in the last four decades would demand that the industry get the most out of that investment until it becomes impossible to miniaturize a silicon device any more. The combination of these two realities would put a conservative timeline of another ten to fifteen years for alternatives to silicon to see the marketplace. Similarly, routine use of nanocomposites in aviation and automotive sectors is also a long-term perspective because these are risk-averse industries. Although research on composites has been around for three to four decades, only about ten years ago did aircraft manufacturers start using composites at a small level in constructing planes. As with alternatives to silicon, the use of nanocomposites will also take at least a decade or more.

In all cases, experience with previous technology waves indicates that the society will see benefits in the long term that have not been forecasted or even speculated currently.

Meyya Meyyappan is Chief Scientist for Exploration at the Center for Nanotechnology, NASA Ames Research Center in Moffett Field, CA. Until June 2006, he served as the Director of the Center for Nanotechnology, as well as a Senior Scientist. He is a founding member of the Interagency Working Group on Nanotechnology (IWGN) established by the Office of Science and Technology Policy (OSTP). The IWGN is responsible for putting together the National Nanotechnology Initiative.

Dr. Meyyappan is a Fellow of the Institute of Electrical and Electronics Engineers (IEEE), the Electrochemical Society (ECS), and the California Council of Science and Technology. He is currently the President of the IEEE's Nanotechnology Council (2006-2007). He is the IEEE Distinguished Lecturer on Nanotechnology and ASME's Distinguished Lecturer on Nanotechnology. He has received many awards for his work and leadership in nanotechnology including NASA's Outstanding Leadership Medal and the President's Meritorious Award. Dr. Meyyappan has authored or co-authored over 150 articles and made more than 200 presentations in nanotechnology subjects across the world.

Clearly, we cannot address all of these issues in depth. In this chapter, we will focus our attention on the ethical issues raised by impacts on the workforce and investment of public dollars.

6.3   Clarifying Purpose

Initially, the purpose of considering the social and ethical implications of nanotechnology can be summarized with the decision points discussed in an earlier chapter.

- Do we fund?
- Do we transfer knowledge?
- Do we regulate?

The first question gets at the intended goals and purposes of nanoscience and nanotechnology. What should be funded and are there research areas that shouldn't be funded, either because they are too speculative, too controversial, or pose some realistic and unacceptable risk at this point in time? The second question raises the issues that might cause us to hesitate in transferring knowledge to other economic sectors or to other nations. On the other hand, this second question could also identify issues that might compel us to share scientific discoveries openly and promote a more rapid transfer of knowledge. For example, should certain military technologies be released to the commercial sector where they could become a threat in the wrong hands? Should medical technologies intended to cure disease or repair injuries be released for use in applications intended to enhance human functional abilities? Alternatively, should technologies capable of alleviating widespread suffering in developing countries be openly shared? The third question addresses concerns that might warrant control or oversight by a third party charged with safeguarding the public interest, while also allowing the marketplace to remain competitive. Regulation can create obstacles to innovation, but the process of slowing down targeted developments may also help guide new technologies in directions more acceptable to society at large.[9] An example here might be the consideration of stronger safeguards on personal privacy before allowing nano-enabled sensors and tracking devices into commercial production, or even a total moratorium on certain classes of nanomaterials if found to be particularly toxic or persistent in the environment.

Beyond these basic decisions, which should be embedded throughout the research and development process from bench science to market release, are specific issues related to the disruptive nature of nanotechnology and the public trust. Should we be concerned about workplace safety? How will nanotechnology expand or shrink the current job market? How can we prepare for the possibility of major shifts in the global economy? How do we know our substantial investment of public funding is being used effectively in the interest of the public good? How do we ensure that the economic benefits of government investment reach as many people as possible?

6.4 Framing the Ethical Questions

If we are to focus narrowly on workforce issues and public investment, a number of ethical questions can be raised using the three basic principles presented near the end of Chapter Four. However, to do so means we need to examine the underlying assumptions of each principle in a bit more depth (Bennett-Woods 2007; Bennett-Woods 2006; Bennett-Woods and Fisher 2004).[10,11,12]

6.4.1 The Principle of Respect for Communities

The principle of respect for persons requires that each person be treated with dignity and as an autonomous agent. Autonomous agents are capable of informed decision making on the basis of what they consider to be their own best interests. As much as a corporation can be legally designated a person with certain inherent interests and responsibilities, a community can be considered an autonomous, self-governing agent with the ability to make decisions in its own best interests; thus, the principle of respect for communities requires that we respect the ability of communities to act as autonomous, self-governing agents, at least insofar as is possible in a complex society.[12] This principle assumes that communities are capable of some level of considered, autonomous self-governance; they have some shared foundations of communal values and goals; and they have the necessary information and understanding to make informed decisions. This last point requires that those with information in science, business, and government share it transparently and without coercive intent. All three entities occupy privileged positions insofar as they may influence social change in ways that are becoming increasingly rapid, complex, and potentially irreversible. Whether they are willing to adopt this type of moral foundation with respect to communities and issues of science and technology has been questioned (Sarewitz 2003; Visvanathan 2003).[13,14]

While the basic concept of informed consent is applicable, the process is obviously more complex and there are many practical barriers to obtaining such consent. The difficulties in obtaining communal consent are particularly apparent with emerging technologies, beginning with the "compulsory nature of technology"[13] and extending to the variation in how different stakeholders might define communal best interests. Barriers also include the absence of mechanisms for public engagement, lack of information and transparency, and societal barriers, such as poverty and illiteracy. Take, for instance, the relative absence of effective mechanisms for timely public engagement in the development and introduction of new technologies. Even if there were effective forums for public dialogue, informed consent requires that citizens have the information needed to conduct a reasoned deliberation. In a rapidly paced and highly competitive marketplace, the interests of researchers and corporations are not always well served by being too forthcoming about the technical nature of their developments and, in particular, any untoward hazards. Again, even if the information were available, communities marginalized by poverty and illiteracy are not likely to receive it or take advantage of public forums to express their views. Nonetheless, calls for public engagement in nanotechnology are calls for some form of informed consent to moderate technological advance in ways that preserve the dignity and self-determined interests of the larger community, while allowing for beneficial progress to the extent to which it can be predicted and assessed. The NNI's emphasis on research initiatives related to public education and perceptions could be an opportunity to find new and meaningful methods of obtaining community consent.

With this in mind, examples of general questions can be posed as noted in Box 6.1. Specific questions regarding workforce issues and public interests can also be formulated along these lines.

BOX 6.1 PRINCIPLE OF RESPECT FOR COMMUNITIES

Guiding Questions

- How is human dignity served?
- Is there potential to violate fundamental human rights including privacy, freedom of conscience, or other basic liberties?
- Is there sufficient information available regarding potential outcomes for communal informed consent?
- Have we employed due diligence in securing a representative level of communal informed consent?
- Does the consent meet at least minimal standards of competence and voluntariness?
- Have vulnerable populations been identified and are there sufficient protections to ensure that vulnerable populations within larger communities have a meaningful voice?

Source: Bennett-Woods, D. and Fisher, E. (2004). *Nanotechnology and the IRB: Toward a New Paradigm for Analysis and Dialogue.* Paper presented at the Joint meeting of the European Association for the Study of Science and Technology and Society for Social Studies of Science (Paris, France). 2004. Paper available online: http://www.csi.ensmp.fr/csi/4S/index.php

- Is the information regarding nanotechnology, and the manner in which it is being presented to the public, respectful of the public's autonomy in terms of being accurate, complete, unbiased, and free from coercive influence and narrow self-interest?
- Does the community have enough accurate, reliable, and useful information to allow it to participate in decisions regarding the potential for widespread job losses, wage impacts, the need for workforce preparation, the possibility of safety issues in the workplace, or the allotment of research funding in particular economic sectors?
- Have we identified those populations most likely to be exposed to workplace safety concerns or to lose their jobs; and have we provided them with enough information, time, and viable options to respond in the best interests of themselves and their families?

6.4.2 The Principle of the Common Good

Derived from a combination of the ethical principles of utility and beneficence, the principle of the common good obligates us to act in ways that respect shared values and promote the common good of communities.[12] On one level, this involves an obligation to promote the well-being of individuals and groups specifically by avoiding harm, minimizing risks, and maximizing benefits. On another level, the concept of a common good is a bit more complicated, but basically encourages actions that respect those shared values that allow us to live effectively in community. In the communitarian traditions of political science and ethics, defining the common good gives attention to a balance between individual liberty and individual responsibility toward the community, including obligations of sustainability and justice between generations.

Defining the common good is very difficult to pin down in the larger scope of society. The first complicating factor is that, in a complex society, we function as many smaller communities nested within larger ones, often with competing interests and differing values. Second, in order to avoid harms you need to know what those harms are likely to be, something that is only a matter of informed speculation in the case of nanotechnology and other emerging technologies. If a technology results in serious environmental degradation or exposes certain populations to a high level of risk, you could convincingly argue that the common good has been violated. However, it is not always clear how to identify the greater harm when the impacts are more subtle or complex. How much unintended environmental degradation is acceptable if food production is greatly enhanced? Technology is often referred to as a two-edged sword that results in opportunities or solves problems for some, while creating barriers or other problems along the way (Lightman et al. 2003).[15] To complicate matters further, when applying this principle, you also need to agree on how to define and weigh various societal benefits with harms, not just in the short-term but also in the longer term.

When considerations of harm and benefit are applied in simplistic utilitarian assumptions, by focusing on short-term happiness of either the greatest number or the few who stand to gain economically, there is potential to overlook the happiness of those who either dissent, are unable to benefit, or are actively harmed by a particular innovation.

Moving beyond benefits and harms to the context of shared values is even more challenging. What are the shared values and best interests of society and what responsibilities do we owe one another? Where do we draw the line between the ideals of community and individual liberty? In other words, where does my liberty stop and the interests of the larger community kick in? For Etzioni (1995),[16] the paramount assumption is that both individual human dignity and the social, or communal, dimension of human existence be recognized and embodied through values of mutual respect, individual liberty, and civic responsibility. Individual liberty can only be preserved when self-interest is moderated by communal obligations.[16] Reeve (2002),[17] in discussing the social implications of technological development, echoes this assumption in asserting "scientists owe a public justification of their activity transcending immediate self-interest, both within their own community and to the wider society."[17] While it may be impossible to articulate one common set of human values or ideals, common sense dictates that some shared values are necessary to our ability to live effectively in community. Such a list might include cooperation, caring, reciprocity, truth, fairness, justice, protection from unnecessary harm, access to basic human needs (including shelter and sustenance), and personal fulfillment.[12] In other words, the assessment of benefits, harms, and overall risk transcends simplistic, all-or-nothing questions of whether or not to proceed with particular nanotechnology objectives, but must include more contextual questions of how best to accomplish specific human ends.

When considering the principle of the common good, several additional questions (listed in Box 6.2) can be posed. Once again, these can then be applied to our specific issues and, while similar to those posed under the principle of respect for communities, these questions are less about informed participation and more about the extent to which others have met obligations to avoid harm and to reflect the community's values in their actions and choices on the community's behalf.

- What are the most likely outcomes of a nano-driven marketplace? Have unintended outcomes been anticipated and risks adequately defined? Have prudent safeguards been put in place to support the well-being of displaced workers and the health and safety of all other workers?

- Have other valuable avenues of research and development been overlooked, harmed, or simply set aside in order to fund work in nanotechnology, especially work that is highly speculative and very long-term?

BOX 6.2 PRINCIPLE OF THE COMMON GOOD

Guiding Questions

- How are the values and priorities of communities represented and served?

- How might the values and priorities of communities be violated or undermined?

- What are the potential short-, medium-, and long-term benefits and burdens for individuals and communities?

- What are the most likely outcomes—positive, negative, or neutral?

- What unintended outcomes can be anticipated? What is the level of risk?

- What limitations or safeguards are prudent to prevent negative outcomes, including misuse?

Source: Bennett-Woods, D. and E. Fisher. 2004. *Nanotechnology and the IRB: Toward a New Paradigm for Analysis and Dialogue.* Paper presented at the Joint meeting of the European Association for the Study of Science and Technology, and Society for Social Studies of Science. Paris, France. 2004. Paper available online: http://www.csi.ensmp.fr/csi/4S/index.php

---

- Have community values and priorities been represented in decision making regarding the nano-marketplace, workplace safety, and funding priorities? What community values or priorities will be violated or undermined by the decisions made and actions taken?

- Have we identified those populations most likely to be exposed to workplace safety concerns or to lose their jobs, and have we provided particular safeguards and contingencies to prevent them from being further harmed?

The principle of the common good essentially boils down to holding all members of a community responsible for the well-being of the larger society and the interests of individual community members. In the face of rapid or unprecedented technological innovation, effective consideration of the common good can allow for a mutually beneficial transition for all members of the larger society. Rather than being a static barrier to scientific and technological goals, the common good can be viewed as a cooperative, interdependent, and highly discerning partner capable of defining and redefining its own interests and priorities in relation to innovation.

6.4.3 The Principle of Social Justice

A final principle is that of social justice, obligating us to act in ways that maximize the just distribution of benefits and burdens within and among communities.[12] Perhaps more than the first two principles, this principle calls us to think globally. The nature of major technological innovations is that they initially become available largely to those who can afford them. A secondary effect is that related outcomes, such as environmental degradation or political and economic disruptions, tend to result in the least advantaged segments of the global society experiencing the highest level of burden and risk, while receiving little or no benefit. The principle of justice asks us to always give primary consideration to alleviating the suffering of the most vulnerable among us.

The sustainability movement provides an obvious example of just such an opportunity. A particularly distinguishing feature of communitarian thinking is a focus on sustainability in both the short and long terms. For example, whether considering issues of environmental, social, economic, or political sustainability, questions of justice force the close examination of current market practices that accelerate technology assimilation for purposes of short-term profits without due consideration or accountability for longer term outcomes. At one level, society can be seen as a collection of nested communities, all competing for basically the same resources and a fair and adequate share of social goods. Meeting the common needs and aspirations of all current members, with a measure of justice, while also preserving resources, avoiding environmental degradation, and ensuring the well-being of future generations, extends communal boundaries to a global scale and requires a collective dialogue about short- and long-term accountability between the communities of science, private enterprise, government, and the public.

General questions of social justice appear in Box 6.3. Specific questions might include the following. Note that, with this principle, it is particularly important to look beyond the borders of any one country to assess the impact more globally.

- How will the benefits and burdens of nanotechnology and its social impacts be balanced within and between communities? Will current social, economic, and political boundaries be enhanced or disrupted?

- Do we have an obligation to distribute the potential benefits and burdens of nanotechnology fairly? How will those communities that are harmed be compensated? Who is accountable for the fair distribution of benefits?

- Have the potential conflicts of interest inherent in close collaboration between the powerful entities of government, academia, and business resulted in an unjust distribution of the public's resources? Is too much benefit going to too few, without fair consideration of the public's contribution?

Taken together, these three principles strive to balance concerns at the levels of the individual, the local community, and the global society.

> **BOX 6.3 PRINCIPLE OF SOCIAL JUSTICE**
>
> Guiding Questions
>
> - What communities are likely to benefit?
> - What communities are not likely to benefit?
> - What communities are likely to experience burdens?
> - Are one or more vulnerable populations at higher risk than the community at large?
> - How are benefits and burdens balanced across communities and between communities?
> - How might current social, economic, and political boundaries be enhanced or disrupted?
> - How will the balance of benefits and burdens be measured?
> - How will those communities that are harmed be compensated?
> - How will social and economic accountabilities be assessed and established in the event of negative outcomes?
>
> *Source:* Bennett-Woods, D. and E. Fisher. 2004. *Nanotechnology and the IRB: Toward a New Paradigm for Analysis and Dialogue.* Paper presented at the Joint meeting of the European Association for the Study of Science and Technology, and Society for Social Studies of Science. Paris, France. 2004. Paper available online: http://www.csi.ensmp.fr/csi/4S/index.php

6.5 Utilitarian Priorities

Technology is an inherently utilitarian proposition. Technologies that do not result in a positive balance of costs and benefits, that are of no positive use, or that cause more harm than good, should be discarded and replaced with those better able to meet human needs. However, this idealized attitude toward technology is not necessarily reflective of how we actually value specific technologies or how we make decisions regarding our use of technology. There are many examples of harmful technologies that remained in the marketplace for too long, even after the harms had become evident,[18] either because there was no better option, or those who profited from its continued presence were unwilling to remove it. On an individual level, most of us don't get into our car every day and consciously weigh the need to get to work against the impact of our planned commute on global warming. Our decisions are a complex mix of short-term utilitarian factors along with habit, perception, awareness, motivation, and values, among others. Failure to address either the negative impacts or the missed opportunities of technology, in the larger society, are often due to a similar failure to reflect fully on the impact of our decisions, at least until much later and once the harm is widespread or the missed opportunity finally realized.

Decisions about how to invest the public treasury in scientific research and development are largely portrayed as utilitarian in nature. This is evident, for example, in the many documents produced by the NNI.[19] Investments in nanotechnology are portrayed in terms of revolutionary benefits: cheaper, more efficient manufacturing of new and existing consumer products, medical breakthroughs, solutions to environmental problems, enhanced national security, and so on. These benefits are accompanied by often dire warnings of falling behind and losing our global competitive edge, the loss of jobs and markets, and even threats to national security. The message is overwhelmingly that the greater risk is in not moving quickly ahead on all fronts. The utilitarian downside has mostly been lumped together under the category of societal impacts, and basically, with the exception of health and safety concerns, handed over to the social scientists to sort out, with a nod to public engagement while they are at it. I know this sounds cynical, but this is not my intent. There are great potential benefits related to nanotechnology and there are legitimate concerns related to global competition, over-regulation, and lack of public support.[9] However, there are also legitimate concerns regarding the other potential disruptions nanotechnology may cause. The point is that without a more concerted effort to coordinate societal implications considerations with the science and technology itself—as it emerges— the odds are great that the technology will arrive well ahead of any workable resolutions to the more problematic societal concerns.

The insurance industry offers an interesting perspective that is also fundamentally utilitarian in its outlook and method (Hett 2007).[20] Hett suggests that nanotechnology represents a fundamentally new kind of risk that takes us back to an earlier discussion regarding the revolutionary versus evolutionary nature of nanotechnology. Coming down on the side of revolution, Hett describes nanotechnology as having no history of gradual adaptation and refinement, and no evidence of performance over time. Because of the novel properties projected for nano-manufactured products, she characterizes the risks of nanotechnology as "unforeseeable" with "recurrent and cumulative losses" being a distinct possibility. In addition, this possibility is made greater by the likely rapid emergence of nanotechnology in several industry sectors simultaneously.

While acknowledging the risks inherent in any new technology, Hett points to two important factors that must be considered with nanotechnology.

> First, the dangers per se are becoming more difficult to understand. Technical systems are becoming increasingly complex, and their components are constantly being reduced in size. Second, not only is innovation achieved and produced at ever-greater speed, but today's technology and business networks also disseminate their innovation faster and over wider areas.[20]

Taking these factors into consideration is an actuarial necessity for the insurance industry and Hett suggests that nanotechnology represents a substantial risk for loss at this point and advocates a precautionary approach. This position has obvious ramifications for investors, regulators, and corporate leaders who must all weigh the potential risks against the potential benefits of funding research and development, regulating the process, and pushing products to market. But it also has ramifications for the public, whose resources are funding it at the cost of other investments in the public sphere.

In a utilitarian framework, science and technology are morally neutral—neither inherently good nor bad. They can only be ethically judged against their outcomes. Since the outcomes are going to be different in different economic sectors, different time frames, and different communities, the principle of the common good forces us to step back and examine nanotechnology more holistically in terms of its more complex systemic effects. Clear priorities should be given to those aspects of nanotechnology most likely to result in good to the community with the least amount of harm or cost. To do so, such priorities must reflect more than narrowly economic and market-based values.

6.6 Conflicting Duties

The question of workplace safety brings to mind the ongoing conflict of duties we face when multiple loyalties are present. In a recent report by the National Institute for Occupational Safety and Health (NIOSH), it was acknowledged that, "as with any new technology, the earliest and most extensive exposures to engineered nanoparticles are most likely to occur in the workplace."[21] The small size and relatively large surface area of engineered nanoparticles may give them unique chemical, physical, and biological properties that pose a danger to the human body. However, little is known yet about what those risks actually are for most nanoparticles, or the extent of exposure in most workplaces. There are few technologies specifically adapted to nanoscale sensing, filtering, and protection. Finally, despite an increasing number of workers involved with research, development, and production of nanomaterials, there is little specific guidance for occupational health surveillance.[21]

So, while there is a clear mandate to investigate worker safety, with obvious efforts being made on the part of NIOSH to do so, the research and development process continues to proceed based on other factors. Designated agents of the government, researchers, and employers have a duty of loyalty to workers to protect them from avoidable harm. However, they also have duties of loyalty to their funding sources, including private investors, to make sufficient progress to retain funding or demonstrate a return on investment. They may even have conflicting loyalties involving consumers, as in the case of medical researchers working on last hope cures or military researchers working on technologies intended to protect soldiers. A well-intended desire to get an application into the field can override caution. Competitive pressures to produce publishable findings in academia or commercially viable products in business and industry can come into conflict when safety concerns threaten to slow down research and development. Excessive regulation and overly cautious safety measures can be costly and can prove to have been unnecessary in the long run. On the other hand, when serious health concerns do arise, employers face the prospect of costly legal action and loss of confidence on the part of investors and consumers. In the absence of definitive information regarding risks, one must balance the potential for harm to workers with the harms of falling behind the discovery curve or arriving late to the marketplace.

One of the problems with regulatory approaches to enforce duties to workers is that the law is essentially a minimum standard. The general attitude toward regulation is often somewhat negative and accompanied by strategies to do only what the law requires in order to be legally compliant. In the face of substantially unknown risks, a better approach is to see one's ethical duty as primary and one's legal duty as secondary. The ethical duty of fidelity or loyalty to workers requires that one take very reasonable precautions against harm and always act in worker's best interest. Loyalty and respect for persons require that one be truthful about the potential risks so that workers can make informed choices about complying with safety standards and guidelines. Truthfulness is also required when untoward exposures occur or new information becomes available that would necessitate medical follow-up. In addition, justice is an issue when not all workers are given the same access to information or provided the same safeguards as others.

6.7   Virtuous Intentions

The power of nanotechnology as an economic force has been credited with the potential to redistribute political and economic power across the globe (Smith 2001).[22] Traditional industries face potentially massive investments in retooling manufacturing processes and large-scale disruption of the workforce. It is likely that certain segments of communities or entire communities will experience a much higher level of burden or risk, while receiving little or no benefit. In anticipation of these burdens and risks, virtue calls us to reflect upon how the motives and intentions behind our actions serve the larger purposes of human excellence and thriving. When we focus narrowly on our own immediate or short-term economic self-interest, without regard to those who depend on us directly or indirectly for their own livelihood and well-being, we fail a certain test of individual and/or organizational character. As nanotechnology unfolds and economic power and opportunity are redistributed across the globe, governments, businesses, scientists, and engineers all have an opportunity to adopt and act upon intentions that balance self-interest with the greater communal interest.

As mentioned in an earlier chapter, the faster the pace, the more complex the situation and the more uncertain its outcome, the greater the need for individuals, organizations, institutions, and society as a whole to be able to work together to earn and retain the public trust. The meaning of a particular technology to human thriving goes beyond its narrow function to include its contribution to individual well-being. Nanotechnologies that have the potential to either cause or alleviate suffering require a commitment of shared character to do what is necessary to develop and distribute them in the best interests of the larger human community.

6.8 Assessing Options for Action

Current efforts to address societal implications can potentially help maximize the benefits of nanotechnology while also addressing its inherent costs and harms. A moratorium on all nanotechnology research and development fails to recognize its potential benefits and the likely inevitability of its development to some level, by someone, as a matter of human nature and need. At the same time, rapid and unencumbered pursuit of any and all potential applications at the nanoscale fails to recognize a shared responsibility for morally conscientious development that is accountable to the larger society. All of the possible steps in between will be most effective if they are clear in their intent and well coordinated from a systems perspective.

6.9 Finding Common Ground

Finding common ground on all the big issues in nanotechnology means finding new and innovative ways to explore and resolve questions of community consent, the common good, and social justice, among others. Technology is not an unqualified good and increasingly represents a force for increased pace, complexity, and uncertainty with which many individuals and communities are struggling. The faster a technology is disseminated and becomes ubiquitous in the larger society, the less autonomy we have as communities to make informed choices. The controversy and public backlash over genetically modified organisms in Europe is generally portrayed in the nanotechnology literature as something to be feared and avoided, a hysterical response to a harmless innovation. Perhaps it is better seen as a reasonable response to the forced introduction of a technology before the community had an opportunity to be informed and to weigh its benefits against communal values and priorities. The actual value of the technology became secondary to the legitimate claims of society to be treated with the basic dignity and respect that should be afforded all autonomous human beings and their respective communities.

Similarly, technological utility must be acknowledged to extend well past a technology's designated function to include concerns such as job security, educational opportunities, workplace safety, and the responsible allocation of scarce resources. Viewing or trying to address each of these issues in relative isolation impedes effective utilitarian analysis in the presence of rapid pace, complexity, and uncertainty. Cost/benefit models that innovatively project and incorporate societal implications in measurable terms will help interested stakeholders better understand the opportunities afforded by nanotechnology now and into the future.

Finally, the question of social justice is paramount to achieving societal support for nanotechnology and its benefits as a means of addressing social issues. Traditional models of endlessly expanding markets, wealth trickling down to everyone, and global prosperity that tolerates a widening gap between the rich and the poor, the powerful and the powerless, feel increasingly unstable and unsustainable. Whether your immediate concern is keeping your manufacturing job and the health insurance it provides; or coping with an inability to feed your family due to a workplace injury for which the employer denies liability; or struggling with a disabling disease whose prior research funding was cut and shifted to nano-projects; or finding clean water in the midst of a developing world slum—the common thread is justice and what is owed in the here and now to those among us who have little or no voice. True common ground has to give meaningful attention to the global priorities of those who are least advantaged and most likely to experience their disenfranchisement in terms of economic loss, environmental degradation, poverty and disease, or assimilation into groups whose only recourse to hopelessness is violence.

6.10   Pragmatic Considerations

> For a successful technology, reality must take precedence over public relations, for Nature cannot be fooled.
>
> **—Richard P. Feynman**

The field of nanotechnology, while suffering from an acknowledged lack of conceptual clarity, would seem to hold great promise for many applications that will serve the needs and desires of society. It also poses real and substantial challenges. True pragmatism advises us to give attention to all of the challenges and seek solutions that prevent concerns from materializing when possible; quickly mitigate harms that are unavoidable; and strategically support those applications most likely to address ethical mandates, such as those suggested in this chapter. Preventing workplace hazards as a matter of business strategy is a better solution than sweeping regulation. Adjusting the pace of innovation and shifting resources into workforce education and training that are adequate to the task is also a good business strategy in the long run. Finally, targeting scarce resources based on some level of public participation, and doing so fairly and with full transparency, is a better strategy for public buy-in than financial arrangements plagued with the potential for abuse of power and conflict of interest.

What then are the practical things that can be done to support nanotechnology's development and introduction to the marketplace that address basic ethical aspects of the process?

Do we fund?

- Involve representatives of the public in developing criteria for prioritizing funding targets and levels.
- Give precedence to funding proposals that make an effort to acknowledge and address societal implications as part of the fabric of the research process.
- Present a more realistic and balanced message to the public about everything from nanotechnology's potential applications and timeline, to its economic impact, to its potential risks. Be prepared to show how these factors affected actual funding decisions.
- Reduce funding for research that does not yet have strong support among the public sector.

Do we transfer knowledge?

- Perform an in-depth and ongoing risk assessment of those nanotechnologies that are most likely to be harmful or misused in contexts other than for what they were developed. If the risks are or become unacceptably high, discontinue or delay funding altogether.
- If continued funding is well justified, then place appropriate safeguards against the transfer of knowledge and create contingency plans for the inevitable accidental, surreptitious, or malicious transfers.

Do we regulate?

- Develop a regulatory model that favors precaution and foresight over post-market reactivity.
- Assure broad stakeholder representation in the design of policy and regulation, and implement safeguards against conflict of interest, politicization, and a limited short-term focus.
- Establish systems that ensure transparency of all research findings, particularly those that suggest untoward or harmful effects.

COMMENTARY

Michael Mehta

"Nanotechnology" is a word that has come a long way in the past few years. Until recently, most people associated nanotechnology with science fiction-based accounts that tended to focus on fantastical devices and applications. Due to recent developments in nanoscience (e.g., greater control over atomic structure and relatively better predictability of nanoscale properties), nanotechnology has entered the commercial realm, and has simultaneously begun the journey of finding its space within the social imaginary.

To become a mature and sustainable technology, nanotechnology must have general support by users of nano-based products and by the public at-large. Since much of the early discussion on nanotechnology focused on how to frame benefits and control risks, it is important to examine the social and ethical impacts of nanotechnology, and the specific challenges arising from the use of this suite of technologies in medicine, industrial processes and products, and food.

As in the case of previous technologies, nanotechnology is outpacing our collective ability to understand and direct its course (Hunt and Mehta 2006). If we consider developments in nuclear technology, information technology, and biotechnology, there appears to be a fairly consistent pattern of development, use, social concern, regulation, and ultimately, some form of resolution. All modern technologies, and perhaps even non-modern ones, move through such a series of stages. This said, it is worth noting that not all technologies survive these transitions. Some technologies, like civilian nuclear power (especially in the United States, Canada, and the UK), and agricultural biotechnology (so-called GM foods) stall in their tracks, and represent case-studies for people in business schools on the topic of commercial failure. With nanotechnology, much is at stake. Since nanotechnology crosses over into so many disciplines, potential and actual business ventures, and is converging strongly with biotechnology in particular, a range of challenges and opportunities emerge.

Scientists have opened up the "black box" of nanotechnology, and as a result, have unleashed a transformative (or disruptive) suite of technologies on the world. Nanotechnology is disruptive in so far as it puts pressure on other products or processes to re-align themselves around its introduction. More importantly, nanotechnology is transformative in the sense that it has the potential, at least in theory, to transform social relations, labor, international economies, and to affect a range of institutions. When we opened up this black box, we ushered in a "nanotechnological" way of seeing the world (Milburn 2002). Consequently, the very existence of nanotechnology plays a role in shaping how we understand the fundamental nature of matter and ultimately affects how we redesign our regulatory, legal, social, and ethical frameworks.

It is crucial to recognize that the success or failure of nanotechnology is contingent on the degree of support it receives from the public. This support can be nurtured by a thorough discussion of the risks and benefits, by encouraging the development of an appropriately constituted regulatory system that can deal with issues like convergence and novelty, and by fostering an environment that seeks public input early and often (Mehta 2005). In short, nanotechnology has moved from the laboratories of the world into the public sphere. Its future depends heavily on how it, and its proponents, navigate and negotiate the world of the social.

References

Hunt, Geoffrey and Michael Mehta, eds. 2006. *Nanotechnology: Risk, Ethics and Law*. London: Earthscan.

Mehta, Michael. 2005. "Regulating biotechnology and nanotechnology in Canada: A post-normal science approach for inclusion of the fourth helix." *International Journal of Contemporary Sociology* 42(1):107-120.

Milburn, Colin. 2002. "Nanotechnology in the age of posthuman engineering: Science fiction as science." *Configurations* 10:261-295.

Reprinted with permission. This article appeared as "Nanotechnology and the public sphere" in the Gradient Corporation's EH&S Nano News, February 2007, Volume 2(2): 1.

Michael Mehta is a Professor of Sociology and Chair of the Sociology of Biotechnology Program at the University of Saskatchewan. He is involved in research on a wide array of topics. Dr. Mehta has helped build a body of literature dealing with genetic testing and new technologies, biotechnology and risk, public perceptions of different applications in biotechnology, and biotechnology and social cohesion. He also explores the emerging connections between biotechnology and nanotechnology, with a special focus on ethics, public policy, and risk. He is the author and editor of five books, with his most recent title being *Nanotechnology: Risk, Ethics and Law* (Earthscan 2006). His website is http://www.policynut.com.

By periodically posing these questions at those key decision points during the nanotechnology development process, and targeting the most questionable applications for more careful scrutiny and planning, potential problems can more effectively be anticipated and solutions developed or applications abandoned. On the other hand, failure to address these issues effectively can lead to widespread public rejection of nano- and other technologies, including those that offer great potential benefits to the society as a whole.

6.11   Questions for Thought

1. Nano-driven advances in business and industry will require a highly trained and specialized workforce that does not currently exist. While this opens the door of opportunity for a segment of the workforce with the means to retool itself, large numbers of manufacturing jobs are likely to be rendered obsolete or displaced. Furthermore, the competition for global dominance in nanotechnology may continue the current trend of new jobs moving offshore, depending on which countries take the scientific and technological lead in nano-innovation. As corporations position themselves to compete in the global marketplace, what obligations, if any, do they have to the local communities disrupted by rapid shifts in the job market, and economic uncertainties? What constitutes an acceptable balance between corporate well-being and individual well-being?

2. The extraordinary investment in nano-research has, by necessity, reduced funding available for unrelated research interests, educational initiatives, and other funding priorities. When faced with scarce resources, every funding decision is a pragmatic choice to be sure; however, every funding decision is also a statement of what the funding source values, and this may or may not align well with the values of the various community stakeholders. What areas of nanotechnology development may not align well with all community stakeholders and what role should the community play in allocating research funding to these areas?

7

Military and National Security Implications of NT

7.1   In Search of a Peaceful Future

> "But for heaven's sake—you're wizards! You can do magic! Surely you
> can sort out—well—anything!
>   Scrimgeour turned slowly on the spot and exchanged an incredulous
> look with Fudge, who really did manage a smile this time as he said
> kindly, 'The trouble is, the other side can do magic too, Prime Minister.'"
>
> —J. K. Rowling[1]

In this scene from the sixth book of the Harry Potter series (Rowling 2005),[1] the current and former Ministers of Magic have come to warn the Muggle Prime Minister that Lord Voldemort has returned and the wizarding community is once again at war. The Muggle leader, awed by the wizards' fantastical powers to conjure items from thin air and travel through the fireplace, momentarily forgets a tragic truth about the ability of humans to wage war on one another. No offensive or defensive strategy can ever guarantee a permanent advantage for the simple reason there is no shortage of new and effective ways in which we can threaten, or be threatened, by one another. Nor is there a foolproof way to keep any technological advantage from eventually making its way into the hands of the enemy. History is littered with the battles waged and lost by the conventionally superior force when it is confronted with new weapons, new tactics, or sufficient resolve on the part of the enemy to wear it down.

In fact, Fudge's admission that the other side "can do magic too" is vaguely reminiscent of the dominant military doctrine of the Cold War. Mutually assured destruction, or MAD, held that so long as each nation possessed nuclear arsenals too large to be taken out in a single preemptive attack, neither side would risk the inevitable retaliatory attack that would devastate, if not annihilate, both nations. The end of the Cold War brought along with it the realization that the aging stockpiles of enough nuclear weapons to destroy human life, as we know it several times over, has become a liability. This threat has become markedly more pronounced as we move into a future that is characterized much less as a battle of will between global superpowers. Rather, the dominant themes in today's conflicts are better characterized by political instability in regionalized conflicts. These regionalized conflicts are largely driven by nationalistic or other ideological ambitions that often transcend political borders, conventional military forces, and even governments themselves.

The 2001 attack on the World Trade Center in New York, and subsequent attacks in Madrid and London, have made it abundantly clear that global security and a workable balance of power will prove at least as elusive in this century as it did in the last. In fact, with these attacks and the rising tide of terrorism, the very context of military conflict and national security has shifted dramatically from that of the 20th century. It should not be surprising then to find that one of the most closely examined areas of nanotechnology research and development is that of military and national security applications. The defense sector has long been a primary supporter of basic research and development. In FY 2007, the Department of Defense (DOD) surpassed all other federal agencies, except the National Science Foundation, with its request for a $345 million investment in nanotechnology research and development, nearly 30% of the entire NNI budget (Roco 2006),[2] along with an estimated $100 million in Congressional earmarks that are outside the NNI plan (NNI 2007).[3] The FY 2008 request increases to $374 million.[3]

Most observers believe advances in nanotechnology are likely to reshape the nature of both warfare and national defense (Altmann 2006; Shipbaugh 2006; Ratner and Ratner 2004).[4,5,6] An appreciation for the nature of these changes is important to understand and assess the ethical implications of nanotechnology and other converging technologies with respect to military and national security issues. In this chapter, we will review certain challenges faced when approaching issues of national security and global conflict, followed by a brief overview of the extensive, potential range of applications of nanoscience and technology in the military and other national security efforts. We will explore the ethical dimensions inherent in assessing both offensive and defensive technologies. However, the ethical considerations of these applied technologies extend well beyond overt warfare and national defense. Technologies intended for military use are also likely to have commercial potential (Altmann 2006),[4] lending additional support to research and development priorities and funding. Likewise, homeland security concerns are not limited to defending against potential terrorist attacks, but include responding to natural disasters, the threat of a global pandemic, the flow of undocumented workers across national borders, and other issues that threaten our domestic safety and security. In addition, a closely related issue of mounting public concern is that of personal privacy, with nano-enabled surveillance capabilities among the most troublesome worries of the general public (Cobb and Macoubrie 2004).[7] Therefore, we will also briefly focus on privacy as a special case of the common good.

7.2 The Context Described: NT and the Art of War

In order to describe the context of nanotechnology and the military, we will begin with a brief examination of the shifting nature of military action and move on to a review of the initiatives currently underway in the military arena. The Institute for Soldier Nanotechnologies (ISN) at MIT is perhaps the most widely recognized research effort to militarize the potential of nanotechnology. We will also visit the somewhat broader goals of the Future Force Warrior program and sample the activities of the Defense Advanced Research Projects Agency (DARPA). Finally, we will summarize the major areas of military applications of nanoscience and technology as currently anticipated.

7.2.1   A Brief Historical Perspective

Rattner and Rattner (2004)[6] describe what they term the new battlespace. One outcome of the strategy of mutually assured destruction was a shift in focus to the developing world of Asia, the Middle East, and Africa. In an effort to avoid direct escalation, both the United States and the Soviet Union began to encourage and support localized guerilla wars by arming and training existing troops, as well as insurgents and militias, in nations that were classified as important to our respective national interests. These now famously include Vietnam, Afghanistan, Cuba, Korea, Iran, and Iraq. It is now accepted that those efforts often undermined legitimate, established governments on the basis of the dominant powers' perceived economic or political self-interests. In retrospect, it is not hard to recognize that this approach sowed the seeds for many of today's most politically volatile situations and much of the rise in terrorist activity across the globe. Unlike the conventional forces faced in earlier world wars, Ratner and Ratner point out that these self-claimed enemies of the West "are trained in unconventional warfare and, like good martial artists, apply force precisely at their enemy's weakest points, often using his own strength against him. This means killing civilians, toppling symbols and monuments, disrupting the economy, and where possible, using the other side's own technology, as happened on 9/11."[6]

In this new battlespace, the authors make the intriguing assertion that victory is no longer the driving factor it was in earlier global conflagrations, where high rates of casualties, both military and civilian, were tolerated along with massive destruction of infrastructure. They attribute this to a psychological shift that began with the wars in Korea and Vietnam. They assert that, by and large, we are less inclined to tolerate human loss, less likely to hold citizens responsible for brutal and irresponsible leaders, and more likely to question both our involvements in distant conflicts and our role in creating the conditions of those conflicts. On the other hand, today's terrorist enemy appears to be well-funded and well-armed, yet unconventional in appearance, tactics, weapons, and their willingness to wreck maximum military and civilian casualties in any location. The specter of weapons of mass destruction, including nuclear, biological, and chemical, in the hands of individuals and regimes bent on destruction for the sake of ideology rather than pragmatic self-interest, is a real and present danger.

Finally, and perhaps even more importantly, we have come to recognize the unacceptable economic price of war. In a globalized economy, the penalties are far-reaching as conflict ripples across closely linked financial markets and trade zones, limiting international travel, weakening consumer spending, and decreasing investment. As noted by Ratner and Ratner, "unlike in the days of empire, war no longer pays."[6]

It is in this volatile and shifting climate we turn to the myriad of possibilities nanotechnology presents. The breadth of possibilities has been well addressed by Altmann (2006)[4] and others, so it should be acknowledged that the brief overview presented here, as a foundation for discussion, fails to do justice to the true potential under consideration and study.

Unlike our experience of nuclear arms, in which THE BOMB invoked frightening images of mass destruction and large-scale devastation resulting from a single, short-lived event, nanotechnology offers a much more varied and nuanced set of military opportunities. While hard-line defenders of nuclear deterrence will argue its defensive role, nuclear weapons are primarily offensive in nature with the image of the mushroom cloud vividly embedded in our collective understanding of nuclear war. In contrast, many nano-enabled technologies have both direct offensive and defensive potentials, rendering a clear characterization of nanotechnology's military and security roles more difficult.

7.2.2   Major Initiatives

The Institute for Soldier Nanotechnologies (ISN) was launched at the Massachusetts Institute of Technology in 2002 with an initial $50 million grant over five years.[8] The Institute is currently in its second five-year contract. Its mission is to "develop and exploit nanotechnology to dramatically improve the survivability of soldiers."[8] The primary focus is on the development of a high-tech battlesuit that is as lightweight as spandex but also bullet-resistant and capable of automatic communication, health status monitoring, injury response, and advanced protection from chemical and biological agents.

There are five strategic research areas currently being pursued by the ISN. The first area deals with lightweight, multifunctional nano-structured fibers and materials that can serve as the foundation for clothing and gear. The second area is termed "battlesuit medicine." Examples of goals listed on the ISN website include polymers for splinting wounds and stabilizing neck and head injuries, devices for controlled release of medications, and remote diagnostics. The third and fourth research areas are focused respectively on protection from blasts and ballistics and detection of hazardous material threats. The final area focuses on systems integration and communications challenges.

The ISN Institute at MIT works closely with the Future Force Warrior (FFW) program of the U.S. Army Natick Soldier Research, Development and Engineering Center. With a similar but broader focus than the ISN, this program seeks "to create a lightweight, overwhelmingly lethal, fully integrated individual combat system, including weapon, head-to-toe individual protection, netted communications, soldier worn power sources, and enhanced human performance. The program is aimed at providing unsurpassed individual and squad lethality, survivability, communications, and responsiveness—a formidable warrior in an invincible team."[9]

Finally, nanotechnology figures into many projects supported by the Defense Advanced Research Projects Agency (DARPA). DARPA's purpose and mode of operation is to shortcut the traditional research and development process by funding unconventional, and even risky, research initiatives that have the potential to quickly bridge the gap between basic discovery and application. DARPA is structured so as to have maximum autonomy and flexibility with a minimum of bureaucratic oversight, allowing it great latitude in the nature and scope of the projects funded. DARPA basically funds the research that others find too unlikely, too risky, or too controversial.

The projects found on the DARPA website are, indeed, the stuff of science fiction novels.[10] Although the term "nanotechnology" is not prominent on the DARPA website, the nature of the projects clearly includes work on the nanoscale. Micro fuel cells, self-assembling batteries, bio-molecular motors, self-decontaminating surfaces, a vaccination against severe pain, and prosthetic limbs that respond directly to neural signals are examples of current research initiatives.

7.2.3 Summary of Anticipated Military Applications

The programs described in the prior section represent a few of the more widely recognized initiatives; however, military goals for nanotechnology are very broad. Altmann (2006)[4] has done an exhaustive review of likely military applications and their relative time frames. In addition, he has classified them as constituting either a modest, significant, or radical advance. Briefly, Altmann predicts nano-enabled applications in the following areas:

- Smaller, faster, and more capable electronics including computers and communication devices that are likely to be embedded in nearly all components—from uniforms, to weapons, to self-configuring networks of micro-sensors, body implants, and small robots.
- Advanced computer modeling and virtual reality-based training.
- Increasingly sophisticated software and artificial intelligence capabilities, including natural language communication, sensory perception, and autonomous decision making.
- Materials with enhanced properties, such as reduced flammability, permeability, and weight; increased hardness, strength, electri-

cal conductivity, and magnetism; and the ability to self-heal, and change shape or color.

- More efficient energy sources and energy storage devices.
- More efficient engines and other propulsion systems, as well as lighter, faster, more agile, and longer-range vehicles—both manned and unmanned.
- Faster-reacting and more energy dense propellants and explosives.
- Distributed sensors as small as a dust particle that could detect light, sound, seismic activity, magnetic signals, and chemical or biological agents.
- Improved heavy and light armor.
- Lighter and more effective conventional weapons, including small arms and light weapons, including guns with no metal and high density armor-piercing projectiles.
- Soldier systems (a body suit) that interact with the body or enhance its functions through biosensors; fluid collection and testing; automatic drug delivery; biochips for chemical and biological analysis; heating and cooling, wound compression, and splinting; a high strength exoskeleton for lifting and jumping; and devices for data processing and communication.
- Implanted systems for monitoring, data processing, and communication; manipulating brain function; inducing cellular stasis for improved performance and endurance; sensory and neural enhancement; and mechanical enhancement of tissues, such as muscle, bone, and tendon.
- Armed and unarmed autonomous robots and vehicles including mini- and micro-robots used for surveillance and reconnaissance; path finding; sensing of chemical and biological agents; various communications functions; and closed surgical operations within the human body.
- Bio-technical hybrids including augmented animals that have been manipulated to fulfill military tasks or animal organs (e.g., sensory organs) integrated into artificial systems and used for functions similar to those of robots.
- Very small satellites that can be swarmed.
- Advanced nuclear, chemical, and biological weapons, as well as protective applications against chemical and biological agents.[4]

Although most developments are projected in the 5-10 year or 10-20 year range, there are a few areas in which nano-applications are likely to be realized within the next five years. For example, nano-based solar cells and generic distributed sensors could appear in the next five years. New, less sophisticated nano-enabled chemical and biological weapons are likely in 5-10 years as are the initial advances in almost every category. Highly sophisticated applications in artificial intelligence or highly speculative applications with questionable technical feasibility appear in the longer-range projections.[4]

While Altmann does classify a few applications as modest advancements, the vast majority of applications represent significant advancements. Not surprisingly, applications related to soldier systems, implanted systems, autonomous systems, bio-technical hybrids, and biological and chemical weapons are labeled radical, along with certain classes of materials and some advances in computing, artificial intelligence, and electronics.[4]

7.3 Clarifying Purpose

Although brief and non-technical, even this basic overview of the potential for nanotechnology in military applications makes it clear that nanotechnology could completely change the face of the military and how it operates (Shipbaugh 2006).[5] By extension, in a world of increasing geopolitical conflict and instability, how the military operates could completely change the face of our common future. In this light, what are some of the practical problems and decisions against which nanotechnology must be assessed?

One obvious decision is how best to target scarce resources. Money and resources allocated to the military are money and resources not allocated to other societal goods. Another practical problem is defining what exactly our aims are with these myriad military applications. Are they primarily offensive or defensive? Is it our intent to retain unilateral military superiority, and for what end? Yet another practical consideration involves the level of risk we are willing to accept given that any technology we develop is likely to be turned against us at some point in the future. We can pose the related problem of what military technologies to allow into the open markets. Finally, we can ask the pragmatic question of whether or not these technologies will ultimately enable war, making it easier to use military interventions as a primary means of resolving conflict.

7.4 Framing the Ethical Questions

Naturally, war itself raises a broad range of ethical questions that extend well beyond the boundaries of this discussion. In an effort to target our discussion more narrowly on the role of nanotechnology in the military, the following ethical questions could be posed.

- Which investments in military nanotechnologies will provide the greatest benefit at the least cost? On what basis do we weigh the costs of the military against other societal needs?
- Are the pursuits of both offensive and defensive applications equally justifiable from the moral standpoints of utility, duty, justice, and virtue?
- When we place our soldiers in harm's way, do we have a duty of fidelity to develop and provide every available means to ensure their safety, enhance their survivability, and secure their success on the battlefield? Alternatively, do some military applications constitute an unacceptable risk of harm to the environment, to the global balance of power, or to our basic human dignity?
- Do we have similar duties to society of fidelity, nonmaleficence, beneficence, and justice not to develop military applications that can eventually pose a serious threat to national or global peace and security when in the "wrong hands?" If so, what constitutes a morally unacceptable level of threat or risk?
- Is it our intent that, by making soldiers safer, enhancing survivability, and improving the odds of success on the battlefield, at least in the short-term, that we can more easily justify war as an ethically acceptable and expedient means of resolving human conflict?

7.5   A Special Case of the Common Good

Recall that the principle of the common good extends the more narrowly defined definitions of risk and benefit, which tend to be limited to utilitarian technical and environmental assessments. Rather than focusing on purely utilitarian ends, such as defeating the enemy, the principle of the common good also calls for the assessment of outcomes that either support or violate generally held social and cultural values. It also promotes a broader definition of societal risks and benefits in the short-, medium-, and long-terms (Bennett-Woods and Fisher 2004).[11] Accordingly, this principle considers nanotechnology and the military in terms of how it either does or does not conform to the broadly held values of society and its priorities.

When allocating scarce resources, few would argue against national defense as a justifiable priority. The concept of self-defense is, in fact, easily justified in almost any moral framework, as well as in legal frameworks. A larger question lies in how we consider military investments in light of other pressing social needs, such as health care, education, infrastructure, and so forth, and ultimately how we value those competing needs in the larger society and culture. All of these domains require investment if we are to maintain and improve our collective quality of life. For example, education and health care are critical underpinnings of economic prosperity, and education plays a key role in maintaining a functioning democracy as well. Alternatively, peace and stability are also highly valued, with the military as one important dimension of achieving those ends. In utilitarian terms of risks and benefits, one can question which is the more effective response to the various external threats we face now and in the future: a military force that has aggressively modernized in terms of all of the potential nanotechnologies discussed earlier, or a shifting of investment into other means of resolving conflict and other pressing societal needs? What middle ground might there be in which we strategically maximize both, thus satisfying the moral requirement that we use scarce resources to create the greatest overall good at the least cost or harm and in consideration of the priorities of the community?

This last point begins to get at the question of the moral dimensions of offensive and defensive applications. War creates tremendous human suffering through death, injury, and displacement. War is inherently destabilizing, with a potential for escalation into other countries and regions of the world. War is expensive. It requires tremendous up-front investment in building and maintaining a military capability, as well as the staggering expense of fighting the war itself. For example, the *Washington Post* reports that, as of August 2007, the cost of the Iraq war exceeds $3 billion a week with the total direct cost of the war in excess of $330 billion since its start.[12] These figures do not include the hidden costs of war, such as economic costs related to the price of oil and other commodities, performance of global stock markets and investor confidence, military and civilian contractor casualties, long-term medical care and disability for severe injuries and mental illness, missed opportunities, and loss of global good will (Bilmes and Stiglitz 2006).[13] While a war economy also generates jobs and demand for services, it seems unlikely that such benefits exceed the costs. Therefore, it follows that avoiding war and/or minimizing its impact would be consistent with the common good in most instances. However, given the likelihood that war will persist as a means of conflict into the future, we are still left with the question of whether the development of offensive or defensive capabilities will have the greater effect to discourage or diminish its impact.

Shipbaugh (2006) argues that defensive capabilities, such as ubiquitous sensing systems, could limit the chances of attack while hardening of structures and improvements in personal safety could reduce the impact of attacks and the need for direct retaliation. Defensive strategies may also effectively counter the unlawful warfare associated with biological or chemical weapons. Offensive capabilities that improve our reconnaissance, surveillance, and targeting of weapons could greatly reduce civilian casualties. On the other hand, those same capabilities can encourage attacks if they increase confidence in the attack's success.[5] In fact, Shipbaugh expresses great concern that, "if nanotechnology develops into very new capabilities over the next several decades, then its relatively unrestrained use in war could create horrific situations."[5] Specific concerns include the creation of weapons that can target specific populations, weapons that can survive in the environment for long periods of time, and combat sensor networks used after a conflict to invade privacy and consolidate political power.[5] Finally, Shipbaugh poses the prospect of nanotechnology providing one actor with a potential advantage for which no affordable countermeasure exists.[5] Are we poised at the cusp of an arms race likely to rival, and perhaps surpass, that of nuclear weapons in the last century?[4] Is it in the interest of our common good to do everything possible to avoid a scenario in which "the other side" may gain the ultimate advantage first or at least be the ones willing to use it aggressively?

Defensive capabilities also raise the question of whether we have an obligation to use whatever technology is at our disposal to safeguard the lives and well-being of soldiers. Simple respect for persons, fidelity, and gratitude would clearly argue that we should. Nanotechnologies that minimize injuries and enhance survivability would seem to meet this ethical obligation. However, we are currently experiencing the result of enhanced survivability due to existing improvements in protective equipment and battlefield medicine. Relatively large numbers of soldiers are coming home with severe brain injuries, amputations, and other disabilities for which medical costs are extraordinarily high, and only limited support and services are available once they leave the military. Therefore, defensive capabilities come with much longer-term societal consequences to soldiers, their families, and their communities that have not as yet been well addressed in either practical or ethical terms.

Finally, this leads us to perhaps the most complex of the questions to be considered. Will nanotechnology ultimately make it easier to go to war? If nano-enabled capacities reduce the cost and lower the risk of war, would war come to be valued as a more acceptable means of achieving the common good? What are the long-term consequences of an increasingly militarized approach to global conflict, and does this represent the values and priorities to which we, as a society, do or should aspire?

7.6   Assessing Options for Action

It is presently very unclear which nanotechnology applications will materialize, so efforts to limit research and development are speculative. One approach would be to proceed with all potential applications and capabilities on the grounds that, if an application is feasible, someone will eventually obtain it and so all research and development becomes essentially preemptive in nature. Another approach is to proceed with virtually unlimited research and development with a clear intention to define possible limitations or restrictions as capabilities become technically feasible and better understood. In both of these options, research and development proceed without meaningful questioning of the long-term value or consequences of the knowledge generated. In other words, the science and technology are assumed to be inherently worth pursuing for military applications, following the general downstream model of technology assessment most commonly used now and in the past.

A second approach is that advocated by Shipbaugh,[5] which involves identifying the relative risks and benefits (pros and cons), between defensive and offensive applications, followed by the development of guidelines that seek to balance development in favor of a stable and secure defense. Examples of defensive technologies that might promote international cooperation and peace include non-lethal weapons, protective materials, and warning sensors. Examples of offensive technologies that could aggravate international conflict and war include better delivery platforms, highly targeted and easily deployed lethal weapons, and target detection sensors.[5]

A third, although not mutually exclusive, approach is preventive arms control, which specifically seeks to prevent new military technologies that are considered likely to pose a particular danger from being developed and deployed in the first place (Altmann 2006).[4] In this approach, nanotechnologies would be subject to prospective analysis of the science and their military-operational aspects, assessment based on specific criteria, and efforts to devise possible limits and verification methods. An example of broad criteria that have been proposed include adhering to, and extending, current arms control, disarmament, and international laws; maintaining and improving stability; and protecting humans, the environment, and society (Neuneck and Molling as cited in Altmann).[4] Based on these criteria, Altmann concludes that the technologies that raise the strongest concerns and that justify calling for immediate limitations are:

- distributed sensors;
- new conventional weapons;
- implanted systems and body/manipulation;
- armed autonomous systems;
- mini-/micro-robots with and without weapons functions;
- small satellites and launchers; and
- new chemical and biological weapons.[4]

These and other options that involve assessment of specific functions against some type of risk/benefit guidelines or criteria assume that objective agreement can be reached on what those guidelines or criteria should be, and that parties will then adhere to them. An obvious barrier is that compliance will become harder and harder to verify as nano-enabled technologies become increasingly small and powerful.

Other options include strict limitations on military research and development to primarily defensive capabilities or a total moratorium on nano-enabled military technologies. Agreement to such limitations or a total moratorium would seem unlikely on a global scale at this point in time.

COMMENTARY

Military Nanotechnology: Ban the Most Dangerous Applications Preventively

Jürgen Altmann

Military use of new technology is special. Different from the civilian sector where many efforts are made to prevent damage, the military goal is to develop new possibilities for selective or massive destruction, and to do so as fast as possible. This follows from the very task of the military—winning in armed conflict, where even a slight technological edge can mean a decisive advantage. Because many countries follow such a course, the consequence on the international level is a virtually unlimited arms race, in many cases with higher mutual threat, shorter reaction times, less stability, and reduced security for all. This so-called security dilemma can be contained by mutual limitation of armaments with appropriate verification of compliance. In the Cold War, strategic nuclear weapons as well as ballistic-missile defenses were limited between the United States and the Soviet Union. Biological weapons were prohibited by an international Convention. After 1990, chemical weapons were banned, too, as were nuclear test explosions.

Nanotechnology promises fundamental revolutions in many areas—and the military sector will form no exception. Some applications would be possible without nanotechnology, but nanotechnology converging with other technologies will greatly enhance the possibilities, in particular of small systems and components. Are there reasons to preventively ban certain military uses of nanotechnology?

Armed forces could use nanotechnology in practically all areas, from guidance systems in munitions via variable camouflage to tailored explosives. In faster computers and higher-strength materials, military developments will go in parallel to civilian ones. A few military developments could have general benefits. For example, small, cheap sensors for chemical or biological warfare agents could make verification of the respective prohibitions easier and improve warning of terrorist attacks. However, there are several potential military applications of nanotechnology that would bring strong dangers. Arms-control agreements could be endangered: new, selective agents using nanotechnological advances in biomedicine would remove an important barrier against biological weapons. Autonomous combat systems would jeopardize the international law of warfare, because for a long time they would not be able to discriminate between combatants and non-combatants. Meeting each other at short distance in a crisis would mean strong pressures for fast action. Instability would also follow from small robots covertly deployed inside an opponent's military systems, capable of striking any time. Small, scatterable sensors could be used for invasions into privacy. Micro-aircraft and micro-missiles would provide tools for terrorists. Non-medical body manipulation applied to soldiers before a broad societal debate could produce a precedent that would be difficult to turn back.[1]

At present, by far, the most military research and development of nanotechnology is done in the United States: it covers 80 to 90 percent of the worldwide spending. One-quarter to one-third of the National Nanotechnology Initiative funding is going to the Pentagon (2005: $352 million of $1200 million). Other countries are starting their own military NT programs. For the time being, these are at a much smaller scale, but they are bound to increase if no limits are agreed upon. This includes Russia and China, competent in nanotechnology as well as military high technology. In Israel and India, high-level politicians have argued for breakthroughs in military nanotechnology. Action-reaction mechanisms so well known from the Cold War are beginning to be felt, including a misrepresentation in the United States of Chinese military writing about NT.[2] One does not need to stretch one's imagination to foresee all kinds of nanotechnology-based military systems offered by many exporting countries in 15-25 years.

An ethically responsible approach needs to set out with the goal of international security, not of national military strength or even global dominance.[3] Prevention of large-scale war should be the highest priority, in particular because of the risk of escalation to nuclear war. Nuclear disarmament should be reinvigorated, space weapons should be banned, conventional-force reductions should be extended beyond Europe. With respect to nanotechnology applications, the most dangerous applications should be prevented—not tied to nanotechnology as such, but focusing on the respective military mission, from small missiles to autonomous combat systems. Such limits will not endanger a limited military capability that may still be needed for peace operations.

References

1. If so-called molecular nanotechnology were to arrive, with self-replicating nanorobots, full control over cell processes, superhuman artificial intelligence, etc., extremely stronger dangers from military application would ensue.
2. An article presenting a U.S. RAND study to a Chinese military audience was presented as Chinese plans to attack the United States with nanorobots.
3. The goal of national security, if seen in an enlightened concept, should lead to the same recommendations.

Jürgen Altmann studied physics and did a doctoral dissertation on laser radar at the University of Hamburg. Since 1985, he has studied scientific-technical problems of disarmament, first concerning high-energy laser weapons, then European ballistic-missile defense. In 1988 he founded the Bochum Verification Project (Ruhr-University Bochum, Germany), which does research into the potential of automatic sensor systems for cooperative verification of disarmament and peace agreements. Prospective assessment of new military technologies and analysis of preventive arms-control measures form another focus of his work, including non-lethal weapons and the interactions between civilian and military technologies in aviation research and development. In recent years, J. Altmann has studied military uses of microsystems technologies and then nanotechnology, with a view towards preventive arms-control. His latest project has analyzed potential new, physics-based technologies for non-lethal weapons at the University of Dortmund. He is a co-founder of the German Research Association for Science, Disarmament, and International Security (FONAS) and a deputy speaker of the Committee on Physics and Disarmament of the German Physical Society (DPG).

7.7   Finding Common Ground

No amount of moral reflection or ideological justification can make war anything more than it is—a tragic reality of the past and present human condition. War is responsible for unspeakable human suffering, environmental devastation, the diversion of vast amounts of human and material resources away from other social goods, and ongoing political and economic instability that largely affect those who are already among the planet's most vulnerable populations. The various wars of the 20th century were unprecedented in human history for the sheer level of destruction inflicted on the globe at large. Sources vary widely over the number of people killed as a direct result of 20th century wars, but tend to range between 150 and 185 million. The costs to those survivors maimed, displaced, impoverished, or otherwise touched by the losses of war are beyond calculation.

This reality does not detract from the sacrifice, courage, or skill of those in the military to whose commitment to our defense we owe much of our prosperity and personal freedoms. So long as we tolerate war as a means of resolving conflict, and in the absence of a globally accepted alternative, the military serves a critical role in national and global defense. Sovereign nations can and should be able to act in their self-interest.

And, to be sure, war has its positive outcomes. The systemic impacts of military conflict are like any significant disruption in a complex adaptive system. The adaptive response may well yield social and technological innovation, increased resiliency, and unanticipated benefits, such as new alliances or enhanced insights into strategic relationships and priorities. Dictators and totalitarian regimes are sometimes ousted; reformist and democratically inclined governments may rise; newfound freedoms may evolve; larger geopolitical goals may be achieved; postwar economic opportunities may be stimulated; and the military/industrial complex, as an economic engine in and of itself, is well compensated.

Nonetheless, the price of warfare is steep and the apparent escalation of terrorism and overt military conflict in the current century is not a comforting trend. Weapons of mass destruction, in combination with extreme sociopolitical ideologies of any bent, are an immediate threat, not just to our way of life, but to human life itself on a large scale. All indications are that nanotechnology will ultimately influence the nature of modern and future military conflict. So, we are faced not so much with the question of whether to use nanotechnology in strengthening our military capability, but rather how best to do so.

Simple utilitarian analysis is not adequate in the face of overriding uncertainty. As Hunt (2006) points out, nanotechnologies "may indirectly, by trickling significant efficiency gains into the intensification of non-sustainable socioeconomic relations, push the world into more resource-driven conflict and war over shrinking supplies of oil, water, arable land, and minerals."[14] At the same time, he suggests the alternative, that nanotechnologies "could have an impact on forcing adjustments, compromises, and accommodations in the existing balance of military power that may contain war and large-scale destruction."[14]

In either case, the disruptive potential for military nanotechnologies is great. Finding common ground raises fundamental questions about the acceptability of war, the role of the military in securing and maintaining the peace, and the voice of the larger society in establishing boundaries and priorities. To ethically assess military applications of nanotechnology, we need first to be clear on our motives. The principle of the common good weighs not just the objective and measurable harms and benefits of a decision, but the underlying values as well. An ethical assessment of military nanotechnology, in light of the common good, requires that we clearly define our short- and long-term goals and intentions, set ethical boundaries, and commit to some level of societal transparency and oversight throughout the development process.

7.8   The Context Described: NT and National Security

While there is a great deal of crossover between nanotechnology, the military and issues related to homeland security, there are also issues that are either unique to homeland security or that we might view a bit differently from the vantage point of security here at home. For example, many generic applications of nanotechnology developed by, and for, the military will have the same or similar uses in the civilian marketplace, including enhanced computing and software capabilities, more efficient energy storage and conversion, nearly all medical applications, and robotics.[4]

Ratner and Ratner point to a wide range of issues that are currently viewed as threats to the security of the homeland including global warming, natural disasters, emerging diseases, and access to energy supplies.[6] Nanotechnology has the potential to address many of these issues, including better fabrication of materials to withstand weather extremes and other natural threats; remediation of the environment; better drugs; and cheaper, more efficient vehicles. Distributed sensors and other surveillance applications may be used to monitor air quality, track a pandemic virus through the atmosphere, or detect terrorist threats, such as those from explosives and biological or chemical weapons. On the other hand, there are also applications that have the potential for misuse or harm in the civilian sector. For example, the fabrication of guns and other small arms without metal, along with the capability of ever smaller and more powerful explosives, could easily defeat current efforts to screen for terror attacks in public venues, and become a further barrier to efforts by law enforcement to control weapons in the hands of the general public.

An area of particular concern has been raised regarding privacy.[4,5,6] Military applications of nanotechnology, such as those in computing, communications, and sensor technologies, could greatly enable civilian tracking and surveillance for any number of purposes. Advanced computing and artificial intelligence capabilities, along with embedded communication devices and sensors, enable the collection and mining of vast amounts of personal data, including consumption, patterns, internet and voice monitoring, real time product tracing, location tracking, genetic profiling, and other highly intrusive activities.[6] Of course, concerns regarding the collection and use of financial, medical, and other personal data are already prevalent; however, nanotechnology is likely to take these to a new level (Mehta 2002).[15] Finally, the potential for implantable brain devices, intended for communication and monitoring of brain/body activity and functioning in the military, opens tremendous concerns for actually observing "thought patterns" and potentially manipulating thoughts as well. That such devices could be so small as to be undetectable adds to the overall "yuck" factor.

In a review of studies related to public awareness and attitudes toward nanotechnology, Flagg (2005) concludes that "of risks described for the public, the most feared is loss of personal privacy."[16] In fact, at least two studies found privacy to be of greater concern than military applications that might lead to a nano-enabled arms race (Scheufele and Lewenstein 2005; Cobb and Macoubrie 2004).[17,18]

7.9 Clarifying Purpose

While the transfer of many nanotechnologies to the civilian sector will be unproblematic, others may pose serious threats. The practical problem then becomes how best to prevent the misuse of military applications in the government and commercial sectors. Our purpose here is to focus exclusively on concerns about how to draw the limits between acceptable and unacceptable nano-enabled public surveillance activities for purposes of law enforcement and national security.

7.10 Framing the Ethical Questions

At least three ethical questions can be raised with respect to the specific use of nano-enabled surveillance, tracking, and monitoring functions for domestic purposes.

- Does personal privacy constitute a basic human right? If so, are there exceptions or limits?

- Do we have a duty of respect for persons and human dignity to prevent the use of certain nanotechologies developed for military purposes from finding their way into use by the government to monitor its own citizens? Alternatively, do citizens have an obligation of fidelity to submit to tracking and monitoring as a means of serving a greater good of safety and security? If so, what constitutes a morally unacceptable threshold of threat or risk to justify loss of privacy? Who should define and be held accountable for the determination of risk?

- Will the use of nano-enabled surveillance, tracking, and monitoring capabilities create more harm than benefit?

7.11   A Special Case of the Common Good Revisited

> Without careful consideration, debate, and possibly, new policy action,
> "Big Brother" may end up being very, very small.

—Jacob Heller and Christine Peterson[19]

Privacy is an offshoot of the ethical principle of respect for persons and personal autonomy. The dignity of the human person lies, in part, in our ability to be self-governing and self-determining. This includes the ability to make informed decisions about what other people know about us and under what circumstances they are allowed to gain that information. This ability to control information about ourselves is the basis of the notion of personal privacy. Several ethical principles are further evident in the notion of privacy. We have a strong obligation to avoid the needless harm that can occur to a person when personal information is divulged under the wrong circumstances. There are expectations of fidelity and veracity implicit in our relationships with employers, health care providers, law enforcement agencies, and other societal institutions that information, deemed confidential or legally private, will not be released to third parties or used for purposes other than those for which it was initially intended. When we cannot trust such authorities with information, we are less likely to provide it, which then undermines the ability of those parties to do their jobs effectively. The principle of justice is also an issue, since the accumulation of personal information is subject to an abuse of power that can unfairly target certain individuals or populations on the basis of religion, culture, and other personal characteristics.

There is a general trend toward increasing and codifying specific legal rights to privacy in the United States, as evidenced by the appearance of various legislative and regulatory initiatives protecting personal information. Examples include the Family Educational Rights and Privacy Act (FERPA) in education and the Health Insurance Portability and Accountability Act of 1996 (HIPAA) in health care.[20] However, such rights are not unlimited. Educational and health information can be shared among the appropriate parties when in the best interests of the student or patient. For example, child abuse reporting statutes require that teachers, counselors, and health care professionals report suspected child abuse to the appropriate authorities. Personal information can also be shared with law enforcement under certain circumstances, if there is reason to believe a crime has been committed or the person in question is believed to be an imminent threat to others. It is here that we begin to see the fuzzy line nanotechnology will begin to draw between the personal and the public good.

While personal privacy is a fiercely held value in the United States, so are personal and public safety and security. Domestic terrorism, pandemic disease, and illegal immigration are all examples for which some enhanced level of public surveillance can be justified and public support is likely. Recent controversies over FBI domestic wire-tapping of persons suspected of being linked to terrorist activities demonstrate the dilemmas we will increasingly face as we attempt to balance homeland security with our traditions of personal privacy.[21] In this case, the debate is actually split between the issues of personal privacy itself and the abuse of power when systems intended to provide oversight are thwarted. Proponents of higher levels of public surveillance will claim that if you are not doing anything wrong, there is no need to fear it; however, there is also some basis in fearing that simply being in the wrong place at the wrong time can implicate you in wrongdoing and lead to great personal harm.

The principle of the common good requires that we examine the likely outcomes of our actions, acting on the basis of a greater utility when such utility is clear. The more likely a particular form of surveillance is to safeguard the public, the more easily it can be justified on the basis of a narrowly utilitarian common good. However, this principle also requires us to negotiate and prioritize the limits of our conflicting values. Simple utility is not enough to justify overly intrusive violations of personal privacy. Our basic civil liberties, including freedoms of thought, speech, and association, assume that we will not be intimidated by constant surveillance. The mere presence of some unknown person or entity watching our every move, listening to our conversation, reading our emails, monitoring our financial expenditures, sorting through our medical records, and so forth, is inherently coercive, undermining our dignity and curtailing our autonomy in both overt and subtle ways.

It should be noted that the principles of respect for community and social justice are equally relevant. Communal consent is critical for public acceptance and the ethical use of these technologies in civil society. In terms of social justice, the potential for certain populations to be more vulnerable than others is clearly present, and requires the conscious identification of such vulnerabilities and the intentional development of appropriate safeguards.

7.12 Assessing Options for Action

The options for action on this issue begin with an outright ban on nano-enabled tracking and surveillance devices for any purpose. On the other end of the continuum is an option to allow for full military and commercial development of capacities with targeted regulation after the fact, in cases where harm has been identified or abuse is likely. All options in between require very targeted consideration of specific contexts and varying requirements for oversight and protection, preferably in advance of the technologies becoming available, or at least widely dispersed.

7.13 Finding Common Ground

The value of such capabilities to the military is great enough that it seems unlikely development will not occur. As mentioned earlier, certain tagging and tracking applications, along with large-scale data collection efforts, are already in use in the commercial sector, so the issue goes beyond nanotechnology per se. Widespread support for detection of environmental threats and biological or chemical weapons is also likely. The challenge is the proverbial slippery slope and the compulsory nature of technology itself. Technologies that cannot be easily detected, and that they are rapidly becoming ubiquitous in everyday life may be strongly entrenched before the scope of harm is fully understood and meaningful guidelines or effective legal protections emerge.

On the other hand, privacy may be the American equivalent to Europe's genetically modified organisms. Overreach by government or the commercial sector may trigger a cascade of demands for limitations and safeguards that prevent effective use of nano-enabled capabilities in the legitimate interests of national security. Privacy in the face of nanotechnology is an issue that demands a national/global conversation to balance basic human rights with security. Once again, underlying motives and intentions are the ethical yardstick by which an effective response can be measured. The failure to build in adequate mechanisms of transparency and accountability to the use of these technologies cannot fail but to undermine their moral and ethical legitimacy over time.

7.14 Pragmatic Considerations

The area of military and national security applications of nanotechnology is perhaps the most technically complex in terms of ethical considerations. This is because the context itself is so complex, so uncertain, and so rapid in its evolution. The Pandora's Box of nano-enabled warfare and public surveillance, in a world where national boundaries are shifting and warring parties are no longer under the control of established states, constitutes a near-term threat to all corners of civil society. More than in any other nano-arena, military and national security implications call for immediate international dialogue and citizen involvement.

The sheer range of technologies possible suggests the need to prioritize considerations and perhaps call for a moratorium on research and development in certain areas to allow adequate time and attention to more pressing issues.[4,14] Defensive capabilities, such as threat detection and protective materials,[4] may be the easiest areas in which to reach common ground. Practical and ethical boundaries established for these technologies can then

COMMENTARY

Plenty of Eyes at the Bottom

Chris Toumey

For anyone who wants better information about the world we live in, nanotechnology produces some spectacular devices: sensors that detect specific molecules; medical diagnostics that locate certain cells; and smaller, more powerful computers. Yet, for anyone who fears that their private life can be detected, diagnosed, computerized, or recorded, these same tools are troubling.

Those who create new ways to see into our lives and our bodies need to know about fear of nanotechnology. M. Cobb and J. Macoubrie reported that "losing personal privacy" was the most salient concern about nanotech in a survey done in 2004, while D. Scheufele and B. Lewenstein detected a similar sentiment the next year.[1,2] The upcoming technologies for sensing and surveillance can be described now,[3,4,5] which makes it important to consider today how they will affect us, and how people will react. In the history of technology, reacting after the fact usually means reacting too late to do anything.

Nanotechnology and privacy

If fears about nanotechnology threatening our privacy are framed within preexisting fears of information systems—if nanotech is seen as a malevolent ally of big, thirsty information systems—then nanotech will bear a burden of proof to show that it will *not* harm one's privacy.

Along with big, thirsty information systems, there are plenty of eyes at the bottom, that is, exquisitely precise molecular sensors that will probably be able to detect where we have been by sampling environmental clues on our clothes. These include physical sensors, chemical sensors, and biosensors.[3] A person with asthma could have a detector to warn of excess pollen or other dangers. That sounds good, but who else will have that information about where we have been?

A third family of phenomena is the molecularly naked patient. Nanoscale cantilevers and other devices lead to diagnostic information with molecular precision. Insurance companies will know much more about our bodies than we will. If genetic discrimination is an established problem, then nano-enabled diagnostics has the potential to intensify it. And if nano-enabled diagnostics is tantamount to loss of privacy, then people may feel they have to choose between privacy and medical care.

Such a choice is sad and ironic. One of the more optimistic themes of nanomedicine is the potential for personalized medicine: highly sensitive diagnostics will lead to highly sensitive therapies. But this means that diagnostic information about the molecules in one's body must be compared with the same kind of information about other peoples' molecules, and that is done by subjecting personal medical information to large information systems.

Finally, there is fear of mechanical miniaturization. Most nanotech insiders consider the bottom-up assembly of atoms and molecules to be sexier than the top-down reduction of larger objects. In the popular imagination, however, nanotech is a top-down process.[2] Without denying that mechanical miniaturization is happening, this means that molecular sensors will be underestimated by a large part of the population that does not think in terms of atoms, molecules, and nanometers.

Privacy-friendly nanotechnologies

Let us imagine that not everything will be grim. There can be ways for nanotech to protect or enhance one's privacy.

First possibility: can nanomedicine create consumer diagnostic tools that do not have to interface with information systems? There was a time when a woman had to go to a doctor to find out whether she was pregnant. Then, the invention of the home pregnancy test kit enabled a woman to discover her status without sharing that information with anyone else. There are also very good personal devices for diabetics to privately measure their blood glucose levels.

Second possibility: souveillance. *Sur*veillance means to observe from above: states use closed-circuit television cameras, radar guns, wiretaps, and other means to see what its citizens are doing. *Sou*veillance is to see from below: citizens use technology to observe what the state and its agents are doing.[6] Recall the 1991 videotape of the beating of Rodney King by a group of Los Angeles police. That record was used to discredit and prosecute the police.

Third possibility: let the risks be known to the citizen. At a discussion on nanotech and privacy in San Francisco in November 2006, Kathryn Hollar raised an intriguing idea. If tobacco products have health warnings, then other kinds of products could have privacy warnings. RFIDs (radio frequency identifications) and bar-code discount cards can have statements saying, "Caution: this product may be harmful to your privacy."

Fourth possibility: appreciate that some nano-enabled surveillance tools are beneficial if the citizen understands the tool and chooses it freely. There are locators, for example, that find Alzheimer's patients when they wander away.[7] There is also great value in nanomedical diagnostics, especially if the patient's informed consent is really informed.

Finally, when people take things personally, some will react personally. It is my hope that threats to our privacy will have to face existential heroes like Rosa Parks, who can personalize and crystallize the problems, and perhaps show us the way to solutions. In the meantime, however, there is not a lot of optimistic talk. With big, thirsty information systems, plenty of eyes at the bottom, molecularly naked patients, and mechanical miniaturization, it is hard to imagine how multiple developments in nanotech will *not* intrude further into our privacy.

References

1. Cobb, M. and J. Macoubrie. 2004. *Journal of Nanoparticle Research* 6:395-405.
2. Scheufele, D. and B. Lewenstein. 2005. *Journal of Nanoparticle Research* 7:659-667.
3. Nagel, D. and S. Smith. 2006. In *Nanotechnology: Science, Innovation and Opportunity*, L. Foster, ed. Upper Saddle River NJ: Prentice Hall, 163-167.
4. Li, M., X. Tang, and M. Roukes. 2007. *Nature Nanotechnology* 2:114-120.
5. Mamin, J. 2007. *Nature Nanotechnology* 2:81-82.
6. Hoffman, J. *N.Y. Times Magazine*, 10 December 2006, 74-75.
7. *Nature Nanotechnology* (editorial), 2:1 (2007).
   Condensed, with permission from the author, from an article that appeared in *Nature Nanotechnology* 2:192-193 in April 2007.

Chris Toumey is a cultural anthropologist who earned his Ph.D. at the University of North Carolina. His work on public scientific controversies includes dozens of articles plus two books: *God's Own Scientists* (Rutgers University Press 1994) and *Conjuring Science* (Llewellyn Publications 1996). In 2003, Chris became interested in the question of how non-experts could have active and constructive roles in nanotechnology policy discussions. This led to several articles on the theoretical grounding of that problem, especially "National Discourses on Democratizing Nanotechnology" (*Quaderni*, Fall 2006). Currently, he works at the University of South Carolina NanoCenter, where he pursues questions of societal and ethical interactions with nanotech. In addition, he directs the South Carolina Citizens' School of Nanotechnology (SCCSN), a dialogue-based program in which scientists and non-experts exchange scientific knowledge and social knowledge, i.e., the values and concerns about nanotech.

be expanded as each new area is assessed. Core questions regarding what should be funded, when and how knowledge should be transferred, and exactly what and how should we regulate, can provide the framework for raising critical practical and ethical questions throughout the process. While it may be important to guard against unbounded idealism and empty platitudes, emphasis on the moral focus of the issues should take precedence over the political and economic interests of parties if the goal is a sustainable foundation for future peace and stability.

7.15   Questions for Thought

1. Altmann favors limits on the military development of military technology with special attention to distributed sensors, new conventional weapons, implanted systems and body manipulation, armed autonomous systems, mini-/micro-robots with and without weapon functions, small satellites and launchers, and new chemical and biological weapons. Select one or two of these and construct your own ethical justifications for and against continued development of these nano-enabled applications, the options for action, and your proposed solution.

2. Should we prioritize investments in offensive or defensive applications of nanotechnology? If so, construct criteria that could be used to do so. To what extent, if any, should the public be involved in military planning and innovation?

3. How much privacy are you willing to give up in the interests of national security? Should nanotechnology be used for domestic surveillance and, if so, in what forms and for what specific purposes? What safeguards, if any, would need to be in place for you to feel comfortable with nano-enabled surveillance? Construct your own ethical justifications for and against continued development of nano-enabled surveillance for both military and civilian security applications, the options for action, and your proposed solution.

8

Sustainability and the Environment

8.1 In Search of a Sustainable Future

> The "control of nature" is a phrase conceived in arrogance, born of the Neanderthal age of biology and philosophy, when it was supposed that nature exists for the convenience of man.

> —Rachel Carson[1]

Nanotechnology's central aim is the ability to manipulate matter at the atomic scale, arguably perhaps the ultimate control of nature. The claims made on behalf of nanotechnology are a laundry list of ways in which nature can be further manipulated to the benefit or convenience of human society: new classes of materials with extraordinary properties, clean energy, clean water, pollution free manufacturing, increased human life span, and even enhancements to the human species itself. Do the goals of nanotechnology constitute a dangerous arrogance and disregard for our place in the natural order, or an inevitable and beneficial leap in our mastery of the world around us? History would suggest a bit of both. Humans have demonstrated a persistent disregard for their direct effect on the natural environment and a lack of appreciation for the complex interaction of effects in the larger system. On the other hand, there is a growing appreciation, embodied in social movements such as those of environmentalism and sustainability, for the finite limits of many resources and the wide-ranging negative impacts humans are having on the globe as a whole. Nanotechnology is in the interesting position of offering a little bit of something to both sides of the equation.

The many effects being attributed to global warming, from the shrinking of the polar ice sheets to crop failures and extreme weather patterns, has riveted global attention on the possibility that humans may soon reach the capacity of the planet to sustain us, at least as we currently exist. Science fiction accounts have long offered vivid descriptions of future Earth as wasteland of nuclear war, massive environmental pollution, or simple overcrowding. In fact, Amazon.com has an entire genre of science fiction entitled Environmental Catastrophe.[2] This rich literary tradition of environmental disaster can easily fuel the fears of those who view nanotechnology as posing a similar threat to that created by nuclear power plants and nuclear weapons. Others see nanotechnology as the answer to many of the looming environmental threats we currently face.

In either case, the search for a sustainable future goes beyond simply curtailing overt environmental degradation. The environmental justice movement is the result of a growing recognition that environmental problems tend to disproportionately affect the poor, creating additional barriers to progress on related social problems. References to a "nano-divide" address the potential for nanotechnology to accelerate the growing inequities in the current global economy. In this chapter, we will pay particular attention to the intersection of the environmental impacts of nanotechnology and larger issues of global sustainability.

8.2   The Context Described: Sustainability's Promise and Peril

> The system of nature, of which man is a part, tends to be self-balancing, self-adjusting, self-cleansing. Not so with technology.
>
> **—E.F. Schumacher**[3]

Nanotechnology is currently being critiqued in a larger context of increasing alarm about the state of the planet and the widening economic and social divide between the "haves" and "have nots" of the globe. In order to consider the ethical issues and questions that can be raised, one needs to be generally familiar with this larger context. We will briefly explore the nature of the environmental and sustainability movements, the primary environmental and health concerns raised by nanomaterials, and the potential benefits of nanotechnology in relation to select environmental and social concerns on the global level.

8.2.1   Environmentalism and Sustainability

The roots of modern environmentalism are typically traced back to the nineteenth century and influences such as European romanticism and early utilitarian efforts to manage natural resources (Smith 2001).[4] While romanticism expressed itself in an aesthetic protectionism, a more utilitarian conservationism arose from industrial concerns that natural resources be scientifically managed to promote and sustain economic growth.

Four distinct periods of environmentalism can be identified in the United States (Lester 1998).[5] The conservation movement, from 1890 to 1920, emphasized the wise development and use of natural resources for the greater good of society. This emphasis shifted slightly during the preservation movement, between 1920 and 1960, to the protection and preservation of habitat for wildlife and recreation. The environmental movement, beginning in the 1960s, represented a shift to more general concerns regarding the quality of the human environment. Integrating both conservationist and preservationist goals, the environmental movement expanded to a broader set of ecosystem concerns, including protection of endangered species, and addressing air and water pollution. Beginning around 1990, the focus expanded yet again to include global issues of sustainable development, environmental justice, biodiversity, deforestation, and global warming.[5]

The environmental justice movement began with the realization and supporting evidence that certain minority populations, particularly people of color, are faced with a disproportionate share of environmental harms, and the accompanying public health and quality of life burdens, due primarily to their lack of access to decision- and policy-making processes (Agyeman, Bullard, and Evans 2003).[6] As a result, environmental justice looks beyond the immediate effects of environmental degradation to underlying social issues that keep certain populations more vulnerable than others.

Sustainability is an even broader concept, encompassing environmental concerns, social justice, and the longer-term goals of sustainable development. Although there is much disagreement on how best to define sustainability, the most commonly cited definition is that of the Brundtland Commission (World Commission on Environment and Development 1987): "sustainable development is development that meets the needs of the present without compromising the ability of future generations to meet their own needs."[7] The strong focus on intergenerational justice encompasses issues that are likely to worsen over time, including the loss of biodiversity, climate change, exposure to endocrine-disrupting chemicals, ozone depletion, and nuclear wastes (McLaren 2003).[8] One central assumption of the sustainability movement is that the current global economy maintains economic growth largely through ecologically unsustainable production systems that exploit both workers and the environment (Faber and McCarthy 2003).[9] In general, proponents of the movement recommend some form of a precautionary approach to risk (COMEST 2005).[10]

The claims made on behalf of nanotechnology's potential to effect rapid technological, economic, and social changes are sweeping in nature. It should not be surprising that proponents of the rapid advance of nanotechnology research and development find themselves in the cross-hairs of these combined movements. The concerns expressed and problems posed by environmentalists and sustainable development advocates are highly systemic and, therefore, compounded by the factors of complexity and uncertainty discussed previously.

8.2.2 Environmental Risks of Nanotechnology

The environmental risks of nanotechnology are largely defined by its novel nature and how little we actually know and understand about materials at the nanoscale (Kimbrell 2007; Clift 2006; Colvin 2003).[11,12,13] Human beings are regularly exposed to naturally occurring nanoparticles, such as viruses, and those that are the byproduct of human-driven combustion. Nanoparticles may enter the human body by breathing them in through the lungs, absorbing them through the skin, or ingesting them through food and drink. Whether any one particular manufactured nanomaterial will pose a health hazard will depend on its size, surface composition, and bioreactivity. Other factors will include the exposure level at which a substance becomes toxic or otherwise harmful and whether or not it passes through the body or accumulates in tissues and organs. Currently, respiratory and dermal exposures are believed to raise the most likely concerns, especially in the work environment (Sweet and Strohm 2006).[14] Cellular and neural effects tend to be those most often mentioned in current reviews of the literature (Blackwelder 2007; Clift 2006; Sweet and Strohm 2006).[12,14,15] Preliminary studies have been both reassuring and concerning. For example, sunscreens using nanoparticles of titanium dioxide have been on the market for some time and there is evidence that the particles cannot penetrate otherwise healthy skin to a depth that would allow absorption into the body.[14] On the other hand, several recent studies have shown that ultrafine particles inhaled through the lungs can migrate to other tissues and organs, and can concentrate and cause an inflammatory response.[12,14] There is also evidence that some nanoparticles are taken into cells, where they may or may not disrupt cell functioning or result in a toxic response.[12,14] For example, it has been shown that wear and tear on the coatings of joint implants can produce particles that are toxic to cells and can result in bone loss.[13]

Broader impacts on the environment are even more uncertain and will depend, in part, on the extent to which nanomaterials persist in the environment and, subsequently, the extent to which they may bioaccumulate in plants and animals.[12] There are also questions regarding how nanomaterials will degrade under natural conditions and in the presence of other substances with which they may ultimately interact in the environment.[13] The release of nanomaterials into the environment becomes more likely with the increase in waste streams from research and production processes, normal weathering and aging, and disposal or recycling processes. These effects are already well known with materials such as asbestos and cadmium.14 As more products are manufactured and placed in general circulation, the chances increase that engineered nanomaterials will accumulate in soil and groundwater.[13] There is reason to be concerned that nanoparticles could interfere with the metabolism of microorganisms, such as those found in soils; however, this is just one example among all areas involving nano-engineered materials for which no meaningful risk assessment has been made.[12] In the extreme case, there is even speculation that self-replicating "nanobots" could initiate an ecological domino effect that could devastate the biosphere. Although detractors suggest this scenario is highly unlikely, it continues to surface in serious discussions of nanotechnology risks (Uskokovic 2007).[16]

While many nano-engineered materials are not likely to pose a definite and substantial risk to human health or the environment, there is also not adequate research to demonstrate either safety or hazard conclusively at this time. In fact, one could consider the absence of a common and currently workable framework for risk identification and characterization a risk in and of itself, adding to the overall uncertainty.

8.2.3 Potential Benefits of Nanotechnology for Sustainable Development

Despite concerns over safety, there is also a significant potential for environmental benefit from nanotechnologies. One possible application that may take advantage of unique nano-scale properties, such as increased surface area and absorption, is the use of nanomaterials in cleaning up existing pollutants. Nanofiltration systems may help provide clean drinking water by filtering out pathogenic microorganisms and other contaminants (Singer, Salamanca-Buentello, and Darr 2005; Meridian Institute 2005).[17,18] Nano-enabled production and storage of energy may reduce the consumption of fossil fuels and the use of lower efficiency, high-polluting energy sources.[17,18] Other applications may improve the efficiency of agricultural practices, reducing the application of chemicals and lowering water usage.[17,18] In the longer term, new "bottom-up" style approaches to manufacturing could reduce the production of dangerous byproducts and generally reduce the ecological footprint of human society.[16] Singer, et al. are among nanotechnology proponents that believe nanotechnology holds the key to meeting many of the most pressing needs of the developing world.[17]

8.3 Clarifying Purpose

As applied in the overall context of sustainability, nanotechnology raises issues at two distinct levels. First, there is the level of the natural environment and the possibility of untoward environmental impacts. The threat of nanotoxicity in humans and damage associated with long-term persistence in the natural environment must be weighed against opportunities, such as nano-enabled remediation of current environmental problems, production of clean water, and creation of more efficient or alternative sources of energy. Practical decisions regarding what to pursue and how quickly depend on answers to an array of practical questions including:

- How much information about the environmental, health, and safety effects of nanotechnology is enough?

- How much transparency is needed in relation to safeguarding proprietary discoveries?
- What regulatory and other safeguards must be in place to ensure products and processes are not released into the marketplace without adequate safety testing that includes the examination of cumulative and longer-range risks?
- How much regulation is too much to allow for efficient and cost-effective progress?
- When untoward harms do occur, what processes must be in place to detect and mitigate them?

Sustainability issues provide the second purpose for giving nanotechnology close scrutiny. There is much reasonable speculation that nanotechnology's broadly disruptive impact will weigh heaviest on society's most vulnerable members, raising the risk for increased human suffering along with social and political instability. Practical decisions regarding how to approach these issues include:

- What research might be given priority in terms of its societal benefits?
- How can beneficial technologies be most effectively applied to existing environmental and social problems, especially among vulnerable populations?

8.4   Framing the Ethical Questions

Primary ethical questions related to nano-safety and environmental concerns initially bring us back to the basic utilitarian question of risk. However, when considering the ethical dimensions of risk, other ethical issues are also evident. Citing risk communication research, Hett explains why some risks create greater anxiety or alarm than others.[19] In general, risks create more concern and are less acceptable if they are perceived to cause some hidden or irreversible damage over time; pose particular dangers to children, pregnant women, or future generations; or threaten death. The principle of nonmaleficence—avoid needless harm—is violated when we are subjected to such harms without our knowledge and consent. This is particularly true when we are exposed to harm for what are perceived to be trivial or narrowly self-serving reasons. Recent class action lawsuits against pharmaceutical companies in the United States and import restrictions against Chinese manufacturers due to lead paint on toys are examples of our low tolerance for what we classify as either negligence or the purposeful and unethical exposure of the public to avoidable harm purely for profit.

Likewise, we tend to be more concerned about risks that we perceive to be either involuntary versus voluntary or inescapable regardless of personal precautions. These worries speak directly to the ethical principle of autonomy and our belief and desire that we be informed and able to consent to the risks to which we are likely to be exposed. In a somewhat different vein, we also tend to be more concerned about risks that are poorly understood by science or perceived to originate from an unfamiliar or novel source, especially if that source is man-made. We also tend to be more concerned if we detect contradictory statements from responsible sources. These concerns reflect basic violations of the ethical principle of fidelity, the loyalty we expect from trusted "authorities," including scientists, regulators, and even preferred corporate brands. Because we cannot make every consumer decision based on a full knowledge of the technical information regarding a product or process, we must rely on these authorities to display the ethical principle of veracity and be truthful in their statements regarding safety. When releasing a product or process into the marketplace, we also expect such authorities to act with beneficence, that is, in our best interests and with careful regard for our common welfare. Finally, role fidelity is based on an assumption of competence, so we expect that such authorities are competent to act in our best interests and would inform us if they were not.

When any one or combination of these factors is perceived, consumers are more likely to respond with fear. The basic ethical questions raised here are similar to those addressed in Chapter 6. Have we avoided needless harm? Have we told the truth? Are consumers able to make informed choices? Have we demonstrated fidelity by creating and then meeting, or exceeding, industry standards for safety testing and hazard reporting? Have we acted in the best interests of society as a whole?

There are also indirect risks associated with the societal effects of nanotechnology. Many risks are related to the uneven or inequitable distribution of benefits and burdens, raising questions of economic and social justice on a global scale. Beyond the questions raised in Chapter 6 regarding the safety and security of workers exposed to health hazards or loss of their livelihood, nanotechnology also has the dual potential to improve or worsen conditions of widespread poverty, disease, and environmental degradation. Such potential raises questions regarding duties of fidelity and beneficence to all vulnerable populations as well as to future generations. It also calls us to move beyond the narrow cost/benefit analysis suggested with direct environmental and health risks, to a broader utilitarian consideration of how best to use nanotechnology's potential to serve the greater social good over time.

8.5 The Precautionary Principle Applied

Because of the great potential for benefits that might be derived from nano-enabled technologies, any effort to argue against proceeding with development will have to present a compelling counterargument. One such argument against unconstrained research, development, and market release of nano-enabled applications is found in the reasoning that underlies the precautionary principle. In order to introduce the precautionary principle in more depth, let's consider a somewhat crude analogy based on four scenarios involving a parachute.

In the first scenario, you are offered an opportunity to parachute from an airplane several thousand feet in the air. You've never done anything like this. You instinctively know it is an inherently dangerous thing to do; however, you also know that most people who parachute from airplanes do not die and are not seriously injured. You also know that many people enjoy doing it. Purely on the basis of history, statistics tell you that you are likely to survive. Your odds go up with better equipment, adequate training, proper supervision, favorable weather and ground conditions, your age and the condition of your health, your ability to follow instructions under stressful conditions, etc. Your decision about whether to accept the offer will most likely be based on what you know about the risks, your confidence in the factors mentioned above, and your personal sense of adventure or curiosity (your values). Most people will pass, but many will give it a try.

In the second scenario, you are handed a backpack and told it contains a device made mostly of silk and strong cords designed to allow you to jump from an airplane and glide down to earth. Although no one has ever tried it, much thought and basic science has gone into its design, so theoretically it should work. If it does work, you would contribute greatly to the effort to develop this particular technology, which could prove quite valuable in the future. In this case, even though there is a chance it might work, almost any reasonable person would decline to accept on the basis of the considerable risks and uncertainty. Fortunately for science, there are always a handful of individuals who will agree to jump, regardless of the uncertainty.

In both of these scenarios, the individual choice not to jump is obviously an ethically acceptable one. There is no compelling reason to jump out of an airplane and, in fact, there are compelling reasons not to. No one is directly harmed if you decline. The stakes, although high for you if the jump does not go well, are low in the bigger picture of society. However, let's alter the scenario just a bit. You are already in a small plane and a violent storm has badly disabled the plane's ability to stay in the air or land safely. A fire has broken out and the plane is likely to explode before reaching the ground. You are handed that same backpack and advised that jumping is your best chance to survive. Now there is a compelling reason to jump even if no one has ever tried it before. All of the prior considerations that allowed you to weigh your various risks are mostly irrelevant. The cost/benefit analysis has become very narrow in scope and your choice now becomes one of what you are willing to do to survive.

Finally, let's take this third scenario one step further by adding one more detail. You are asked to secure a small child to your chest in order to save her as well. She cannot possibly survive unless you are willing to jump with her. Now your choice is no longer an individual one, but directly affects the life of someone else. If we now metaphorically substitute nanotechnology for the parachute, we can see that all four scenarios are representative of the continuum of choices posed by nanotechnology, from the trivial applications of convenience or entertainment to profound questions of human suffering and meaning in both the present and future. The addition of the child, as a metaphorical representative of the next generation, adds yet another moral dimension in the form of intergenerational justice.

It should be clear that the precautionary principle is most easily situated in the first two scenarios, where precaution is either based on known or predictable risk factors or a relative absence of uncertainty in combination with two clear choices—to jump or not jump. It is less easily applied to situations in which the impacts are broad and systemically interrelated in complex ways or when technological solutions are needed in the near term. The more complex the interrelationships between desirable developments in one arena, and potentially undesirable developments in another, the harder to tease out the moral obligations. Likewise, the more pressing the problem, the harder it is to advocate for a strongly precautionary approach.

The World Commission on the Ethics of Scientific Knowledge and Technology (COMEST) proposes the following working definition of the precautionary principle:

> When human activities may lead to morally unacceptable harm that is scientifically plausible but uncertain, actions shall be taken to avoid or diminish that harm. Morally unacceptable harm refers to harm to humans or the environment that is:
>
> - Threatening to human life or health, or
> - Serious and effectively irreversible, or
> - Inequitable to present or future generations, or
> - Imposed without adequate consideration of the human rights of those affected.[10]

The precautionary principle is presented as an ethical mandate for the application of a precautionary approach to new technologies that pose some threat of untoward harm. It is applied under circumstances in which the following conditions are met:

- There exist considerable scientific uncertainties;
- There exist scenarios of possible harm that are scientifically reasonable;
- Uncertainties cannot be reduced in the short-term without, at the same time, increasing ignorance of other relevant factors by higher levels of abstraction and idealization;
- The potential harm is sufficiently serious, or even irreversible, for present or future generations or otherwise morally unacceptable;

- There is a need to act now, since effective counteraction later will be
  made significantly more difficult or costly at any later time.[10]

COMEST further identifies the moral values on which the principle is
based. The first is the expectation that individuals, corporations, and political states are ethically responsible for their choices. Responsibility can be
mitigated to some extent by one's efforts to make the most informed choice
possible, to prevent harm within the choice, the decision not to act in the
midst of uncertainty, and the extent to which others share responsibility for
the outcomes. Other values include equity, environmental protection, and
democratic representation in decisions.[10]

Common criticisms of the precautionary approach are that it is fundamentally unscientific (Resnick 2002)[20] and logically inconsistent, overly restrictive, or otherwise unworkable (Hahn and Sunstein 2005).[21] Clift points
out that the objective of risk management and regulation for conventional
chemicals is to eliminate risks to humans and the environment or to reduce
them to "acceptable levels." If the hazards associated with exposure and the
exposure pathways are unknown for nanoparticles, then risk can only be
confined if release is avoided. However, in theory, certain forms of epide-
miological evidence would never develop, since there would be no harmful
effects on the human population or environment if all release is avoided.[12]

While the definition of the precautionary principle presented above does
allow for research to progress on the basis of proportionate risks, so little
is currently known or understood about nanomaterials that the calculation of risk is inherently unreliable and highly uncertain. In the meantime,
there are well-known risks to not proceeding. Emerging diseases with the
potential to become pandemics, terrorism and political extremism, global
warming, intractable poverty, and diminishing supplies of clean water are
all examples of complex, high-stakes problems that require innovative and
potentially risky solutions. Furthermore, each of these problems is likely to
be mitigated, at least in part, by technological solutions that currently do not
exist. Nano-enabled solutions currently appear to hold the greatest promise
of proportionate benefits.

Much of nanotechnology is highly speculative and we have little experience on which to base predictions; however, we do have historical precedents
where materials, such as asbestos, were used too quickly and the negative
effects known, but ignored far too long (Harremoes et al. 2001).[22] By placing
all of our concerns and dialogue in the basket of objective risk assessment, we
miss the larger moral point. Asbestos is a straightforward example of a beneficial material that proved hazardous. The larger ethical failure was in the
absence of a virtuous response to the information, once asbestos was in use,
that it posed a direct risk to both human health and the larger environment.

One can objectively know whether a material is safe or unsafe and still
choose to interpret the risks and benefits in self-interested ways. For example, the most questionable environmental implications of nanotechnology lie
in the highly speculative realm of molecular nanotechnology and scenarios

involving self-replicating robots and other hazards. While acknowledging the potential risks, one could argue for an "active" (meaning choose the least risky option and take responsibility for risks) rather than a "strict" interpretation of the precautionary principle on three grounds. First, nanotechnology may represent the only solution to pressing problems. Second and third, inaction may result in the development and use of molecular nanotechnology by questionable parties and, so it follows, the rest of us would not be in a position to respond effectively to its improper or irresponsible use (Phoenix and Treder 2004).[23]

On the surface, these arguments appear to make sense. However, upon moral consideration they neglect a couple of key points. First, they make an assumption that all "pressing problems" require a solution and that a solution is possible. In fact, in complex adaptive systems, solving one problem often creates another, and sometimes a more difficult problem, locally or elsewhere in the system. The moral considerations of harm and best interests require that our response to problems be proportionate to the harm caused by inaction. While lack of certainty or full understanding is not a sufficient reason to discontinue all scientific endeavors, it may be sufficient in situations where we aren't yet capable of fully characterizing the risk potential on a broader systems level. In other words, we don't know enough to be able to identify the least risky option.

The other point they neglect is the underlying coercive nature of the arguments that "if we don't do it, someone else will," and, by extension, "we certainly can't trust them to use it responsibly." This good guy/bad guy characterization of our choices makes a few untenable assumptions. First is the assertion that "we" will use it responsibly and that others may not. We have not always shown the wisdom to use our science and technology in morally accountable ways in the past, and there is no reason to assume we can be counted on to do so in the future. Second is the assertion that development is inevitable, which then seems to preclude serious efforts to prevent development in the first place, even in the presence of legitimate moral or practical justifications. It also ignores the fact that, if we do develop it, there is no assurance the bad guys won't simply steal it and use it against us anyway. Finally, there is the subtle assertion that just because we understand how to make it, we also know how to mitigate its harmful effects. Nuclear, biological, and chemical weapons are all examples of technologies that we well understand how to develop, but not how to protect the population against on even a small scale. In this light, the proposed application of an "active" form of precaution, one that simply limits development to the proverbial good guys, fails to meet ethical requirements that we avoid harm and ensure the best interests of others. It fails the test of utility insofar as having to control a questionable technology after it is already here is more costly and harmful than preventing its development in the first place. It is oppressively paternalistic in that it also presumes to judge who can and cannot be trusted with development. It fails to provide any test of virtuous intent that would conclusively identify whom should and should not be allowed to participate. Finally, it cultivates fear and distrust of others in order to coerce support.

If there is an ethical mandate for precaution with respect to nanotechnology, then further development and refinement of the precautionary principle is warranted. First, application of a precautionary approach will require development of improved models of risk profiling, tracking, and prediction that take into account disparate effects within systems. Second, such a model would ideally be highly contextual. At the same time, it would overtly recognize that treating the same risks differently in different situations is likely to result in knowledge or technology being developed that is used in both positive and negative ways despite our best efforts to the contrary. Finally, a precautionary principle that sets a minimum standard of meeting narrow, pre-selected, risk thresholds, will fail to account for the frailties of human motivation as well as the broader societal impacts that resist prediction and measurement. In other words, a precautionary principle that is viewed more

COMMENTARY

George A. Kimbrell

Engineered and manufactured nanomaterials enter the natural environment throughout their life cycle: through manufacturing, transportation, use, disposal, or intentional introduction. Many nanomaterial products (such as cosmetics and sunscreens) consist of "free" nanoparticles not fixed in a product matrix which speeds up their interaction in the environment. Once loose in nature, these nanomaterials represent a completely new class of manufactured non-biodegradable pollutants.

Nanomaterials' unique chemical and physical characteristics create foreseeable environmental risks, including potentially toxic interactions or compounds, absorption and/or transportation of pollutants, durability or bioaccumulation, and unprecedented mobility for a manufactured material. Environmental impact studies have raised some red flags, such as carbon fullerenes causing significant lipid peroxidation in the brains of largemouth bass after exposure. However, despite rapid nanomaterial commercialization, many potential risks remain dangerously untested due to the failure to prioritize, and paucity of funding for environmental impact research.

Nanomaterials create immense difficulties for existing environmental protection laws like the Clean Air Act and Clean Water Act. Agencies lack cost-effective mechanisms of detecting, monitoring, measuring, or controlling manufactured nanomaterials, let alone removing them from the environment once they are released. These regulatory frameworks are data-driven and with nanotechnologies there is a general dearth of data; company-created data are considered confidential business information and not released. The risk assessments, oversight triggers, toxicity parameters, and threshold minimums in U.S. environmental law are premised on bulk material parameters (such as a relationship between mass and exposure) that are wholly insufficient for nanomaterials. Finally, these laws lack a life cycle framework and fail to address existing regulatory gaps.

Consider now that bête noire of industry, the precautionary principle. Isn't the argument in its favor *even more powerful* with regard to new technological systems such as nanotechnology, that creates substances that can be fundamentally novel and unpredictable, where long-term health and environmental impacts have not been adequately studied, and where existing oversight mechanisms are inadequate? How about the oversight principles of transparency and the public's right to know?

Whether industry and policy makers will avoid the mistakes of past "wonder" substances remains to be seen, but current agency inertia and industry opposition to oversight does not instill confidence that history will not repeat itself with regard to nanomaterials and the environment.

George Kimbrell is a staff attorney for the International Center for Technology Assessment (CTA), where he works on legal and policy issues related to nanotechnology, biotechnology, and climate change technologies. He drafted the first legal action on the human health and environmental risks of nanotechnology, a legal petition filed with the Food and Drug Administration (FDA) on behalf of a coalition of consumer, health, and environmental organizations in May 2006. Mr. Kimbrell has written several articles on nanotechnology oversight and regularly speaks on that topic. Mr. Kimbrell is also a staff attorney for CTA's sister organization, The Center for Food Safety (CFS), where he works on legal and policy issues surrounding genetically engineered foods and crops, organic standards, and aquaculture. He received his law degree magna cum laude from Lewis and Clark Law School.

as a barrier of regulatory micromanagement than a personal moral imperative may fail to nurture the level of moral accountability needed to effectively
manage the disruptive potential of nanotechnology over time.

8.6   A Special Case of Social Justice

> In recent years it has become increasingly apparent that the issue of
> environmental quality is inextricably linked to that of human equal-
> ity. Wherever in the world environmental despoliation and degradation
> is happening, it is almost always linked to questions of social justice,
> equity, rights, and people's quality of life in its widest sense.
>
> —Agyeman, Bullard, and Evans[6]

Critics of the unrestrained development of nanotechnology echo this general
concern with respect to the creation of a "nano-divide" that furthers the economic, political, and social isolation of countries that do not yet have access
to existing technologies (Mehta and Hunt 2006).[24] Countries with access to
advanced technologies are likely to be the first to harness the power of nanotechnology and will remain more successful as they "will continue to gobble
up intellectual property, dominate globally coordinated marketplaces, and
because they will reap military advantages from these technologies."[24] As a
result, these authors believe that nanotechnology will generally work against
the interests of the developing world.

Others believe that nanotechnology offers the best shot for the developing world to get a leg up. Eight goals, dubbed the Millennium Development
Goals, have been established by the United Nations in an effort to address
the most pressing needs of the developing world. They include extreme
poverty and hunger, education, gender equality, health, environmental sustainability, and global awareness and partnership.[25] In an effort to prioritize
nanotechnology applications on the basis of their potential contribution, an
expert panel was convened to identify the top ten nanotechnologies most
likely to help achieve these goals (Salamanca-Buentello et al. 2005).[26] At the
top of the list appear improved energy storage, production and conversion,
followed by enhanced agricultural production, and water treatment and
remediation. With the exception of improved construction materials, the
remainder of the list are directly related to health and include disease diagnosis and screening, drug delivery systems, food processing and storage, air
pollution and remediation, health monitoring, and vector and pest detection and control. The authors of this study recommend establishing a grand
challenge initiative to address each of these areas. They further recommend
supporting the initiative through collaboration among foundations, nanotechnology initiatives, and the private sectors in both developing and industrialized countries.[26]

The real concern, of course, lies not in the potential for nanotechnology to help mitigate deep-seated social problems, but in the mechanisms of access. For example, at least nine biotechnologies, related to medicine and the environment, have been identified as having the potential to greatly impact health in the developing world (Daar, Thorsteinsdóttir, Martin, Smith, Nast, and Singer 2002).[27] However, there are clear barriers to those technologies becoming available. One barrier is a matter of research priorities. The risk profiles of developed and developing countries differ significantly (WHO 2002).[28] If research priorities allocate the majority of nanotechnology research and development toward the diseases of the developed world, coupled with a westernized approach to medicine, health disparities may actually increase (Bennett-Woods 2006; Invernizzi and Foladori 2006).[29,30] A second barrier involves basic market forces and the issue of access.[29] Most companies engaged in nanotechnology development seek a solid return on their investment. Increasingly, the partnerships between research universities and the private sector that have been promoted by the NNI, have resulted in patenting and restrictive use licensing that directly block access to technologies for humanitarian purposes.[18,29]

The principle of social justice requires we act in ways that maximize the just distribution of benefits and burdens within, and among, communities. In order to do so, we must identify what communities are likely to benefit, what communities are not likely to benefit, what communities may actually be harmed or burdened, and the extent to which benefits and burdens are fairly distributed across, and between, communities. We are also called to identify those populations within communities that are most vulnerable. The nature of the benefits and burdens must also be identified along with a means for measuring or assessing the balance. We need to be able to anticipate how current social, economic, and political boundaries might be enhanced or disrupted, and then develop a means for monitoring and responding to such disruptions. Finally, we must ask how communities and populations that are harmed will be compensated, and who will be accountable for negative social and economic outcomes.

Agyeman et al. argue persuasively that sustainability "is at its very heart a political, rather than a technical, construct."[6] Nanotechnology as a tool for or against social justice is, in and of itself, morally neutral. Its ability to serve the interests of social justice lies in the practical and ethical motives and intentions of its creators, regulators, and distributors.

8.7 Assessing Options for Action

One obvious option for action is to ban, or dramatically forestall, further research into nanotechnology until sufficient controls are in place to guarantee its safety and minimize its untoward outcomes. This option seems an unlikely choice for many reasons, including sound moral reasons. From a practical standpoint, we still don't know what controls are needed, so our ability to preemptively put them in place is quite limited. Further research and development in basic nanoscience is needed in order to effect the parallel development of measurement and testing methods. Ethically, the potential benefits of nanotechnology would argue for pursuing at least some applications on the basis of utility and a common or greater good.

The action called for most widely, with respect to environmental, health, and safety issues, is the adoption of a strategic research framework that makes maximum use of national and international, interdisciplinary collaboration to keep up with the pace of discovery and the breadth of nanoscale applications. Such a framework would likely include the development of new measurement techniques, screening tests for toxicity, predictive capabilities for risk evaluation, and approaches to life cycle assessment (Maynard 2007).[31] Life cycle assessment is a proactive approach to managing risks that tracks a product from the beginning to the end of its life cycle.[14]

The Nanoscale Science, Engineering and Technology Subcommittee (NSET) of the National Science and Technology Council (NNI 2006)[32] outlines five current areas of research needed to adequately assess environmental health and safety implications of nanotechnology. The first area deals with instrumentation, measurement, and analytic methods along with the lack of a common terminology and standards for nanotechnology. The area of nanomaterials and human health calls for investigation into the absorption and transport of nanomaterials throughout the body from various oral, inhalational, dermal, and intravenous exposure routes; the nanoscale properties that elicit a biological response; and appropriate quantification, characterization, and prediction models. With respect to the environment, the report calls for research to better understand how nanoscale materials enter, persist, degrade, and are transported through the environment. A fourth area dealing with environmental surveillance focuses on monitoring protocols to determine both the presence of nanomaterials and health outcomes in exposed populations. Finally, work needs to be done in the fifth area of risk management methods with an emphasis on the full product life cycle.[32]

Aside from technical and methodological challenges, such research will be costly and, as noted in an earlier chapter, there are calls for a significant increase in federal funding for this effort. NSET acknowledges that new developments are advancing rapidly and recommends an initial prioritization scheme based on various risk factors, collaboration, and adaptive management that allows for quick changes in direction when called for. At the time of this writing, subsequent refinement of this prioritization scheme is available for public comment (NNI 2007).[33] What is notable regarding these documents is the comprehensive and transparent nature of the process along with the strong emphasis on collaboration between public and private sectors on both a national and international level. The question that remains to be answered is whether or not it is possible to construct options for action to address these proposed research objectives in a manner that can keep up with the pace of new developments in nanoscale research and development. Therefore, this option for action, focused on risk management, leaves open the potential need to slow down at least some areas of research, such as those involving known or reasonably predictable hazards and those characterized by the highest levels of complexity and uncertainty.[16]

A third option is an open-ended approach that simply lets the market sort it out as nanotechnology unfolds. This option will determine the commercial potential of nanotechnologies the most quickly and avoid what may turn out to be unnecessary regulation and front-end investment in testing and monitoring. This option also carries the highest risk for both untoward environmental, health, and safety effects and other undesirable societal outcomes.

With respect to issues of sustainable development and social justice, there appear to be few initiatives and little consensus on what the options are.[26] All options for action that seriously address the disruptive effects of nanotechnology will involve providing nanotechnology solutions, or means of production to developing countries, while also avoiding or mitigating against the potential harms. Simplistic solutions call for simply giving away nanotechnology to the developing world; however, this approach could quickly degrade into making those populations the de facto testing ground for nanotechnology's safety and effectiveness. The idea presented in an earlier chapter that all research, on some level, is human subjects research is relevant here. Viable options for action that best address the ethical imperatives of social justice will be those that promote incentives for the diffusion of nanotechnologies into applications that benefit vulnerable populations, while also safeguarding the population from untoward harm. As noted earlier, the more pressing the problem, the more risk that can be tolerated; however, moral intentions should always be to anticipate and avoid such risks with the same care one might take toward oneself, one's own family, or one's own immediate community.

8.8 Finding Common Ground

> The most important and urgent problems of the technology of today are no longer the satisfactions of the primary needs or of archetypal wishes, but the reparation of the evils and damages by the technology of yesterday.
>
> —Dennis Gabor[34]

As illustrated in the earlier parachute scenarios, the initial, logical, and instinctive response to any potential or perceived threat is to avoid it. However, upon further reflection, this is obviously not the most prudent approach, especially in the face of an even larger or longer-term threat. The key to finding common ground may be in better defining and prioritizing such threats.

COMMENTARY

**Nanotechnology and the Environmental Protection
Agency: Societal and Other Implications**

Nora Savage, Ph.D., and Anita Street, M.P.H.

Nanotechnology is generally defined as the ability to create and use
materials, devices and systems less than 100 nanometers in size, and to
manipulate and view materials at this scale (see http://www.nano.gov
for more information). Nanoscale particles possess distinctive qualities: quantum mechanical effects may be dominant, the surface area to
mass ratio is dramatically increased, and materials may exhibit phys-
iochemical properties quite different from their macro-sized elemental
equivalents. The unique properties of nanomaterials can be used to
enhance and improve a wide range of products and processes—everything from cosmetics, to sports equipment, to car coatings. Consequently, the field of nanotechnology is expected to grow worldwide
with both corporate and governmental entities engaging in the research
and development of nano-based products. Currently, approximately
450 consumer products on the market purportedly contain engineered
nanomaterials. The National Science Foundation forecasts that $1 tril-
lion worth of nano-enabled products will be on the market by 2015.

This rapid surge in worldwide research and development and the
potential large-scale use of nanomaterials in products points toward
the need for clearer information regarding environmental health and
safety impacts. The capacity for nanomaterials to accumulate in certain
organs or other components of living systems needs to be explored,
along with the metabolic alteration of these substances and subsequent effects upon living systems. How these materials move from
one media to another, from one organism or ecosystem to another,
and from organisms to the environment, and vice versa, will be critical for understanding and implementing proper manufacture, use, and
recycle/disposal options. In order to effectively assess these impacts, a
full life cycle analysis (analysis of a product from the accumulation of
starting materials to the development, manufacture, use, and eventual
disposal or reuse of the item or portions thereof) of the various constituents and end products must be undertaken.

Environmental concerns posed by engineered nanomaterials include
their potential persistence and toxicity in the environment, the possible
synergistic effects of nanomaterials with contaminants or naturally
occurring compounds in the environment, and questions regarding
their ultimate fate. How reactive such compounds could be in the environment, what compounds are formed during degradation, and where
and how these compounds partition to various environmental and biological media are crucial, yet largely unknown questions.

The U.S. Environmental Protection Agency (EPA) is committed to supporting collaborative research and coordinated activities among its governmental, non-governmental, and international partners. In addition, EPA supports funding ways to use nanotechnology to prevent and remediate pollution, and also bears responsibility for identifying potential hazards and exposures to humans and ecosystems. For EPA, opportunities exist to employ nanotechnology to prevent and solve environmental problems by developing approaches that will foster innovation, and achieve the environmentally sound production, use, reuse, and disposal of nanomaterials. To accomplish its mission, it is incumbent upon EPA to better understand the potential environmental health risks and benefits associated with the applications of nanotechnology. Perhaps the greatest challenge lies in ensuring that, as nanomaterials are developed and used, unintended consequences and risks are minimized or prevented to the extent possible.

Since 2002, EPA has sponsored approximately $12.2 million in research related to the environmental applications of nanomaterials and $17.8 million on the environmental effects of nanomaterials through extramural grants. An additional $2 million has been expended for research conducted in EPA's laboratories. In addition, the Agency has sponsored over $3 million for Small Business Innovation Research projects on nanotechnology.

In 2004, EPA's Science Policy Council (SPC) created an Agency-wide workgroup to examine nanotechnology from an environmental perspective. In February of 2007, the group released a document to examine and clarify issues in anticipation of the significant impacts resulting from the development of nanotechnology, to include research needs. Accordingly, EPA is engaging in a variety of activities to promote this development in the most responsible way possible. These activities include: sponsoring research on the potential environmental applications and potential implications; and coordinating and participating in strategic research planning meetings, conferences, and workshops to address possible societal and environmental impacts of novel technologies.

The term "responsible development" of nanotechnology has often been used by both proponents and critics of this field. The meaning may vary slightly depending upon the source, but generally it denotes benign development, equitable deployment, and harmless end-of-life disposal of materials or products. Careful consideration of the complete life cycle of materials and products will help eliminate or minimize deleterious human health and environmental consequences. Examples such as chlorofluorocarbons and DDT come to mind to illustrate instances where detailed knowledge of the full life cycle of products and materials might have prevented subsequent environmental problems. The research sponsored by the EPA is targeted towards enhancing scientific understanding of the effects of engineered nanomaterials and exploring the development of environmentally benign nanotechnology processes. Additional efforts underway to address full life cycle considerations include: a consortia composed of all stakeholders working to better understand full life cycle impacts and develop methods and approaches to minimize adverse impacts; coordinated sponsored research among various stakeholders; participating in and organizing workshops to communicate current research results and outline future needs; and collaborating with industry to seek optimal solutions for the environmental challenges connected to nano-based manufacturing processes.

"Responsible development" goes beyond the technical arena and is not limited to risk considerations. This process should involve and encourage an open dialogue with all concerned parties about potential risks and benefits. Public engagement and participation in decisions about research and development are likely to become more of an issue in the future when questions arise concerning the equitable access and distribution of beneficial applications. It is equally as important to gain and maintain public trust and support if the societal benefits of nanotechnology are to be fully realized. Therefore, EPA is committed to keeping the public informed of the potential environmental impacts and applications associated with nanomaterial development and use. As an initial step, EPA is developing a dedicated website to provide a venue for public access of EPA-wide information and also to open a dialogue concerning nanotechnology. In addition, EPA's outreach includes organizing and sponsoring sessions at professional society meetings; and speaking at industry, state, and international nanotechnology meetings.

Adding to the complexity of such philosophical and ethical questions concerning equitable distribution and responsible development is the emergence of convergence—or the melding of nanotechnology with biotechnology, information technology, and cognitive technology. Currently, many researchers are developing nano-bio materials with specifically designed properties. For example, the work of Angela Belcher and lab associates at the Massachusetts Institute of Technology (MIT) involves inducing tiny, benign viruses to produce inorganic materials with semiconducting or magnetic properties as a novel method to remediate heavy metals.

The convergence of nano-, bio-, info-, and cognitive sciences will have a major effect on federal agencies. For EPA, this blend may alter the very nature, scope, and method of conducting business as a regulatory agency. The basic principles by which EPA regulates industries, monitors the environment, and seeks to ensure a clean ecosystem and healthy public could undergo major transformations as developed technologies enable novel capabilities that depend on vastly different premises and assessments. Much of this will result directly from technological advances as traditional disciplinary lines and boundaries are blurred through inter-disciplinary research and product development. Others could be the result of evolving EPA priorities and mandates.

Finally, with increased convergence of technologies, concerns regarding the interaction, overlap, and oversight of the regulatory authorities of federal agencies should be considered. Confusion may result over jurisdictional issues for agencies with oversight of similar materials or products. Regulatory purviews may begin to cover identical topical areas and overlook others. Consequently, important issues could be resolved with differing, conflicting conclusions from different agencies, causing confusion among regulated industries and the general public. For example, the purview of the EPA could overlap and intersect with that of the Food and Drug Administration (FDA) or the Consumer Product Safety Commission (CPSC). Continued collaboration and coordination of activities among regulatory agencies will be critical. This will be increasingly important as products with converging technological processes begin to be manufactured on a mass scale and enter the marketplace.

Nora Savage is an Environmental Engineer at the Environmental Protection Agency (EPA) in Washington, D.C. in the Office of Research and Development. Her focus areas include nanotechnology, pollution prevention, and life cycle approaches for emerging technologies. She is one of the Agency representatives on the Nanoscale Science, Engineering and Technology subcommittee of the National Science and Technology Council and serves as the Vice Chair of a Technical Coordinating Committee in the Air & Waste Management Association. She has authored or co-authored several articles on nanotechnology in leading journals.

Currently, she serves as the lead for the EPA's Office of Research and Development Nanotechnology Research Team, which is developing a nanotechnology research strategy. Her primary responsibility is to enable the EPA to continue to protect human health and the environment as nanotechnology and novel engineered nanomaterials continue to evolve and enter the consumer marketplace. Her ultimate goal is to stimulate the formation of collaborations and liaisons between EPA staff and representatives from other federal agencies, academia, and industry to bring about a holistic approach to achieving proactive environmental protection.

 **Anita Street** is an Environmental Scientist with 17 years of EPA experience. In her career, she has worked in the areas of risk assessment, training and communication, environmental justice, and sustainability. For the past five years in the Office of Research and Development, she has managed an Environmental Futures Project. Visit http://www.epa.gov/osp/efuture.htm.

In a narrow economic sense, regulation is often argued to pose a significant threat insofar as it restricts the development, commercialization, and dissemination of new products into the marketplace. As such, it is seen as a barrier to innovation on a global level where nations that regulate earlier or more stringently are at a competitive disadvantage (Mehta and Hunt 2006).[24] While economic ends are clearly valuable in a practical sense, one can question whether they are the most valuable in the ethical dimension; and this is particularly true when short-term economic benefits to the few, lead to negative economic outcomes in the longer term for society as a whole. Justly serving the economic interests of society means that the economic rewards have to be dispersed in some mutually beneficial way to all segments of the society.

In terms of threats, environmental degradation and human safety are much clearer. No one wants toxic materials degrading in their proverbial backyard. Neither does anyone want to read in the morning paper that nanoparticles from their personal care products have been accumulating in their internal organs, weakening cellular metabolism or immune functions, and predisposing them to cancer or brain disorders in later life. Corporations do not want to be held morally and legally liable for large-scale environmental contamination that must be remediated at great cost. Governments do not want to be faced with a global pandemic, originating in the developing world from organisms mutating in, and being transferred to, the human population through unsanitary water supplies and unsafe agricultural practices. Each of these can be considered an avoidable harm. In each case, the harm can best be avoided by proceeding, but with appropriate caution and care. The greatest barrier at this point becomes the varying definitions and approaches to appropriate caution that are evolving in different international circles (Phelps 2007; Hunt 2006; Goldenburg 2006; Masami et al. 2006; Mills 2006).[35,36,37,38,39]

More complex, but equally compelling, are the threats that come from growth that is not sustainable and the instability that arises from deep-seated and persistent social injustice. From an economic perspective, the poor in any part of the world are a market waiting to happen. However, adding unsustainable levels of consumption to the human ecosystem is a short-sighted approach to human well-being. The sustainability movement calls for new economic models that can make use of our emerging technological capacities in ways that reduce human suffering; enhance human potential; honor cultural beliefs and practices; and do so in a way that honors the ethical commitment of intergenerational justice for our children and grandchildren. Once again, there is much room for common ground here if the will exists to pursue it.

8.9 Pragmatic Considerations

As discussed throughout this chapter, there are two general areas of pragmatic concern regarding the environmental impacts of nanotechnology and its contribution to sustainable development. The first deals with direct environmental impact and the second with access to benefits. With respect to the first issue, there is a consistent message in the current literature on environmental, health, and safety issues that there is not yet enough financial support for adequate assessment (Maynard 2007; Balbus, Denison, Florini, and Walsh 2005).[31,40] Having a comprehensive plan for what needs to be done is only helpful if accompanied by the resource funding to do so.

In addition, it seems clear that the responsibility for making sure this happens may fall increasingly to industry itself, both as a matter of self-interest and as a matter of the practical limitation of expecting government to assume sole responsibility.[40] There is a rising public backlash against what has come to be viewed as blatant irresponsibility on the part of corporate interests for ensuring public safety and the public good. In the face of financial scandals, environmental disasters, massive product recalls, and executive compensation packages that contribute to the increasing concentration of wealth among an ever smaller portion of the world's population, the calls for corporate social responsibility are becoming louder and more persistent. When combined with a similar erosion of trust in government, and the increasingly close relationship between academia and private industry, there is reason for the public to question who exactly is safeguarding their best interests. Business, government, and academia have clear moral obligations to avoid harm, act in the best interests of society, be truthful, be competent in their assigned roles, and balance the social benefits and burdens associated with the public's investment and reliance upon nanotechnology's future benefits.

With respect to the second issue, it is also a matter of practical self-interest for all stakeholders, that questions of sustainable development are addressed sooner rather than later. The frank reality is that nanotechnology, at best, could help alleviate many of the intractable problems that bedevil the developing world, but only if there is a conscious, concerted, and strategic effort to do so. At worst, its disruptive potential may aggravate existing problems as a result of destabilizing industries and local means of production; increasing consumption of new consumer goods, untoward environmental degradation, and even greater concentration of wealth and power on a global scale. These latter outcomes are in no one's long-term best interests, so there are solid, pragmatic reasons for not allowing this to happen.

There are no simple answers or solutions. What is called for is a combination of voluntary initiatives and regulatory responses that creatively and strategically address the need for entirely new systems and processes for safeguarding people and the planet as a whole.[41] Such initiatives must be willing to "just say no" or "let's hold up a bit on that," as well as setting priorities for pressing ahead in the most promising areas. As a matter of the common good, there must be a willingness on the part of stakeholders to give up some short-term gains in scientific discovery, profit, and access, in favor of the long-term benefits of sustainable development and social justice. Effective approaches will explicitly honor the moral obligations noted above, by consciously raising ethical questions, such as those posed earlier in the chapter, at every step in the process. Furthermore, it will take a similarly concerted effort to make sure narrowly defined environmental concerns are integrated with core issues of social justice and sustainability in its broad social, economic, political, and environmental sense.

8.10   Questions for Thought

1. Select an area of interest in nanotechnology and brainstorm its likely benefits and risks. Apply the precautionary principle to your analysis and determine the best course of action based on your interpretation of the principle.

2. Even if nanotechnology development proceeded unhindered and many of the benefits were realized in terms of commercial applications, there is no guarantee these applications would reach the people who most need them. Poor countries and vulnerable populations are not poised to purchase solutions to deep-seated problems on an open market. Develop a set of recommendations for how to get nano-enabled solutions to those who most need them, and provide both ethical and practical justification for your plan.

9

Nanotechnology in Health and Medicine

9.1 In Search of a Healthy Future

In health there is liberty.

—Henri-Frederic Amiel[1]

Health is a pervasive and driving force in our lives. Although we frequently don't fully appreciate the value of good health until it has been compromised, our health enables or disables nearly every other aspect of our lives. In addition, we increasingly look to the health care system to help us maintain a level of health that allows us to remain active and engaged in our lives. We view health care as an important social good, some would argue even a basic right, and we are generally supportive of medical research and advances in medical technology (Bennett-Woods 2007).[2]

Therefore, it should not be surprising that research into public attitudes on nanotechnology reveals solid support for nanotechnology in the areas of health and medicine. In a study reported by Macoubrie (2006), the category of major advances in medicine was the highest ranked among the perceived benefits of nanotechnology, with 31% of those surveyed placing it at the top of the list.[3] Interestingly, this was also the area in which 95% of those surveyed did not trust either the government or industry to effectively manage the possible risks of nanotechnology in medical applications. An earlier study also found "new ways to detect and treat human diseases" to be perceived as the most important benefit of nanotechnology (Cobb and Macoubrie 2004).[4]

Nanotechnology and the human body are no strangers to each other. Our bodies basically operate on the basis of naturally occurring molecular machines, structures, and processes that operate on the nanoscale within our cells. For example, proteins are essentially nanomachines that perform a number of assigned tasks. In fact, the human body has been characterized as the ultimate "nano factory" (Vo-Dinh 2007).[5] A truly informed understanding of both normal human functioning and disease hinges on our ability to observe, understand, and eventually diagnose and treat medical disorders at the nanoscale. New nanodevices, such as nanoscale microscopic devices, nanoprobes, nanosensors, and optical tweezers, are allowing scientists to visualize, probe, sense, track, and manipulate biochemical processes at their molecular level, and enabling rapid advances in the field of molecular biology (VoDinh 2007a; Chan 2006).[5,6]

The term *nanomedicine* has been coined in an attempt to capture the interdisciplinary focus of this rapidly evolving field of medicine. Nanotechnology is predicted to play a critical enabling role in the medical field, as it brings together various disciplines including physics, chemistry, molecular biology, health sciences, and engineering (Ebbesen and Jensen 2006).[7] The convergence of nanotechnology with fields including biotechnology, information technology, and cognitive science (NBIC) is expected to result in new and revolutionary science and technology platforms that will affect many, if not all, areas of medicine (Roco 2005).[8] This intersection of technologies has already resulted in an explosion of research and development efforts in an extraordinary array of medical applications (European Technology Platform 2006; Ebbesen and Jensen 2006).[7,9] The NNI has placed a great deal of emphasis on medical applications of nanotechnology, providing slightly over $200 million in funding for fiscal year 2008 for the nanomedicine initiative of the National Institute of Health's (NIH) Roadmap for Medical Research (NNI 2007).[10]

While much societal implications emphasis has been placed on the technical capabilities of nanomedicine, and especially on the areas of radical human performance enhancement and life extension, relatively little attention has been paid to the more general impact nanomedicine may have on the health care system. In this chapter, we will examine ethical issues related to prioritization of research, cost, access, and the implications of personalized medicine. The more radical developments in human performance enhancement and life extension will be addressed in Chapter 10.

9.2   The Context Described: Nanotechnology and Personalized Medicine

> In nothing do men more nearly approach the gods than in giving health to men.
>
> —**Marcus Tullius Cicero**[11]

Although few nano-based medical products are currently in clinical use, there have been early successes in the development of diagnostic devices, contrast agents, analytical tools, therapeutics, and drug-delivery vehicles.[6] In this section, we will briefly review the major areas of current research and development in nanomedicine, the concept of personalized medicine, and key factors in the current U.S. and global health care environments that relate to societal impacts and potential ethical issues.

9.2.1 Pharmaceuticals and Therapeutics

The pharmaceutical industry has been quick to recognize the tremendous opportunities to develop methods that simplify, speed up, and reduce the costs of drug development and testing, as well as increasing drug safety and efficacy (Ferrari and Downing 2005).[12] The term *targeted drug delivery* refers to one of the most promising and highly anticipated areas of research. It involves the development of nano-enabled medicines that can deliver drugs to specific locations at specific times in a controlled release mode. Work is currently being done on the development of *nanocarriers* that enhance the stability of a drug and its ability to permeate natural barriers, such as membranes within the body (Singer et al. 2007; Alonso 2004).[13,14] Such delivery systems may enable alternative routes of administration for existing drugs that will minimize drug degradation, reduce side effects, and allow site-targeting (Kubik, Bogunia-Kubik, and Sugisaka 2005).[15] Targeting of therapeutic agents is of particular concern in cancer treatments, and this may be where the earliest commercial successes emerge within the next five to ten years (Bullis 2006).[16] For example, biodegradable polymer nanoparticles appear to be ideal candidates for cancer therapy, as well as vaccine delivery, contraceptives, and antibiotics (Kayser, Lemke, and Hernandez-Trejo 2005).[17] Future developments include "smart" delivery systems that are increasingly sensitive and responsive to changes in drug concentration and other factors.

The fields of pharmacogenomics and pharmacogenetics will use nano-enabled applications to speed up drug discovery and enhance the ability to target more effective medicines to the individual patient (Lindpainter 2002).[18] An example might be the development of genetic immunizations using DNA-based vaccines. Nanoparticles may even eventually be used to replace or repair defective or non-functional genes.

9.2.2 Diagnostics and Imaging

New nano-enabled approaches to diagnostics are being investigated, including optical probes based on quantum dots, biological sensors, and improved contrast agents for imaging (Vo-Dinh 2007b).[19,6] Recently developed nanoparticles are able to visualize and measure cellular events during therapy, and may lead eventually to the development of smart nanoparticles that are able to sense and detect the state of biological systems and living organisms optically, electrically, and magnetically (Choi and Baker 2007).[20]

Better image resolution, longer tissue retention, and tissue-specific targeting are being achieved through the use of nanoparticles in contrast agents for medical imaging (Mazzola 2003).[21] Nano-enabled fabrication of high throughput screening microarrays of DNA, protein, carbohydrates, cells, and tissue will enhance rapid genotyping, genetic analysis, and DNA resequencing. Microarrays are also being developed for application in early diagnosis and monitoring of cancer, genetic epidemiology, tissue typing, microbial identification in infectious disease, and drug validation (Campo and Bruce 2005).[22]

9.2.3 Nanoscale Surgery

Nanosurgical instruments and techniques, such as nanoneedles and femto-second laser surgery, are likely to replace current surgical methods. Following on the heels of microsurgery, nanosurgery has already enabled surgical intervention at the level of the individual living cell.[7] Applications are predicted to include cell therapy, eye surgery, neurosurgery, tissue engineering, laser-assisted in-vitro fertilization (IVF), and gene therapy.[7]

9.2.4   Implants and Tissue Engineering

Biocompatibility is a primary barrier in the development of synthetic materials that can be implanted in the human body to replace tissues or organs that have been damaged by disease, injury, or simple wear and tear. Nanomaterials, developed on molecular platforms, make use of unique properties at the nanoscale, such as the ability to mimic the surface dimensions of a protein. These materials may increase biocompatibility and therefore enhance the degree to which medical devices, prostheses, and engineered tissues are able to operate for extended time periods within the body without being rejected (van den Beucken, Walboomers, and Jansen 2005).[7,23] For example, polymer matrices can be used as a biodegradable substrate for the proliferation of replacement tissues derived from stem cells.[7]

Virtually every type of tissue and every human organ are current research targets for tissue engineering.[7] Novel therapeutics are being developed using the unique properties at the nanoscale. For example, the polymer-based artificial cells currently being tested for targeted drug delivery have already been employed for some time as a bioabsorbent that can be used to remove toxins and drugs from the blood. They are now being developed to encapsulate other cells in order to prevent rejection when transplanted. Applications include pancreatic islet cells in the treatment of diabetes, liver, and kidney cells, as well as a number of genetically engineered cells and microorganisms (Chang 2005).[24]

9.2.5   Multifunctional Nanodevices

Finally, much has been written about the possible design of multifunctional nanodevices equipped with molecular motors, sensors, and actuators. Such a device could theoretically operate as a self-contained entity to diagnose, treat, and monitor diseases, such as cancer or Alzheimer's disease.[6] These nanorobots, or "nanobots," represent the most futuristic and controversial of the medical applications to date. There are many technical barriers to overcome, including biocompatibility, power, communication, navigation, and removal (Patel et al. 2006).[25] However, it is not the technical barriers that tend to raise concerns, but rather the ideal functional characteristics of nanorobots to overcome the barriers. These include swarm intelligence (distributive intelligence), cooperative behavior (emergent and evolutionary behavior), and self assembly and replication. Popular science fiction accounts, such as that found in Michael Crichton's *Prey* (2002), emphasize these particular characteristics when imagining scenarios of nanobots out of control.[26]

9.2.6 Personalized Medicine

Within the next 10 to 15 years, one of the predicted outcomes of nano-enabled advances in genomics and proteomics is a fundamental shift in the traditional western model of medicine—from diagnosing and treating acute disease once it develops, to an increasingly predictive and preventative model of what has been termed *personalized medicine* (Weston and Hood 2004).[27] The term, personalized medicine, is a construct that has not yet been well-defined, but generally refers to the ability to tailor medications and other treatments to the specific biology of the individual patient (Califf 2004).[28]

For example, the field of regenerative medicine will be enabled by nanotechnology. Regenerative medicine uses the body's existing mechanisms for tissue regeneration and repair to treat chronic disease conditions or recover from injuries. The clinical goals shift from treating symptoms and slowing down a disease process to the use of engineered biomaterials that can stimulate and monitor regenerative processes to actually reverse what would otherwise have been permanent effects (Pison et al. 2006).[29] Although given much positive attention in literature, the actual promise of personalized medicine is much less clear, especially considering the likely level of investment and ongoing cost it might entail.[28]

9.2.7 The Broader Health Care System

While it is clear that nanotechnology has the potential to radically change the way medicine is practiced, very little attention has been given to related and equally important effects on the health care system itself.[2] Radical change in the delivery of medicine will likely lead to the need for similarly radical change in the basic infrastructure of the health care system, including the health care workforce.[9] The overview of current research and targeted applications presented here emphasizes prevention, early diagnosis, and embedded systems for health monitoring and repair which, in turn, suggest less need for acute inpatient hospital care and more technologically complex primary care and outpatient services. Betta and Clulow (2005) describe this as a fundamental shift from the hospital to the laboratory.[30] Even though this trend has been occurring in health care for a couple of decades now, the rapid dissemination of nano-enabled advances may accelerate the movement away from traditional inpatient hospital care; however, the care that remains in the hospital setting will also reflect the new technological realities.[2,9]

Adapting the current infrastructure of health care toward a new primary focus on prevention will require hospitals, alternate delivery and diagnostic settings, and related industries, including the pharmaceutical and medical device industries, to make substantial investments in retooling facilities to

accommodate new equipment and services. There is a related need to prepare the workforce to shift its attention away from acute to chronic care,
and to incorporate the rapid increase in medical knowledge and the impact
of emerging technologies, such as nano-enabled genomics and proteomics,
in the delivery of health care. The impact is likely to introduce new roles
and the need for new areas of expertise that raise issues, similar to those in
manufacturing, of the need to retrain the existing workforce.[2]

All of these changes will entail significant investments of new resources
and additional costs. Although there are claims that nanotechnology will
make certain processes faster and cheaper, such as microarrays for lab testing,[9] the very idea of personalized medicine, coupled with the need for
companies to recover substantial investments in research and development,
suggests higher costs, at least in the initial stages.[2]

The expenditures for health care in the United States totaled $1.9 trillion
in 2004, 16% of the gross domestic product, and more than any other country (National Center for Health Statistics [NCHS] 2006).[31] Despite efforts
since the early 1980s to control costs and create a more efficient and effective
delivery system, health care expenditures continue to increase at more than
double the rate of inflation. At the same time, the number of Americans without insurance rose to an all-time high of 47 million, or 15.8% of the population (U.S. Census Bureau 2007).[32] Although some technologies do result in
a net reduction in health care expenditures, most do not; and there are not
currently systems in place to evaluate the cost impact of new technologies
on the larger health care system (Kaiser Family Foundation 2007).[33] Finally,
implicit in the goals of medicine is extending as many lives as possible to
their natural limits. Perhaps the most obvious impact is an ever-increasing
percentage of the population living longer. The elderly are the fastest growing segment of the population worldwide (United Nations 2001)[34] and, in the
United States, the fastest growing demographic is the "oldest old" (persons
85 years old and older) (Hobbs n.d.).[35] The aging of the global population has
been linked to numerous social, economic, and medical problems (Callahan
and Topinkova 1998).[36]

9.3   Clarifying Purpose

When considering the impact of nanotechnology in medicine, the initial
focus is obviously on the safety and efficacy of nano-enabled applications.
Medical applications constitute the most intimate interface between human
beings and nanotechnology, as nanotechnologies are directly and intentionally introduced into the human body. A range of concerns are raised including informed consent for research subjects, product safety, long term health
risks, medical nanoparticles entering the external environment, and the ade-
quacy of current regulatory systems charged with ensuring product safety.

A less visible consideration, although equally important, is the impact nanotechnology will have on the deep infrastructure of health care, including the cost of, and access to, services. The health care system in the United States, as in other countries, is already under mounting pressure to expand services while also reducing costs. Practical problems to address include:

- How will the basic infrastructure of health care be altered in response to emerging technologies?
- Will nano-enabled technologies increase or decrease the overall cost of health care?
- How will limited resources be reallocated and who is most likely to benefit?[2]

9.4 Framing the Ethical Questions

A number of writers have suggested that nanomedicine may raise more complex questions than general medicine, or even biotechnology. However, while the nature and extent of some of the problems may be somewhat novel, the underlying ethical concerns are likely to remain very similar.[7] Nanomedicine will raise many of the same questions of autonomy, nonmaleficence, beneficence, veracity, fidelity, and justice that have long been raised in bioethics.

For example, privacy is a long-standing issue in health care and has become more so with the increased availability of genetic testing and the introduction of computerized patient records. We explored issues of privacy related to nano-surveillance in Chapter 7, and essentially the same concerns can be applied regarding privacy in the biomedical realm. Medical probes and sensors could be easily administered without someone's knowledge or consent, and vast genetic databases are being accumulated that could be used for non-medical purposes.

Likewise, medicine has consistently faced questions regarding patient safety, the benefits and burdens of new technologies, and issues of cost and access. Nanotechnology, in and of itself, does not present particularly unique problems; however, its wide scope, potential impact as an enabling technology, and relatively short time frames may accelerate the need for society to confront difficult questions regarding the goals and priorities of medicine and the larger health care system.

- Will nano-enabled biomedical sensors, probes, and genetic testing and monitoring capacities, developed for medical purposes, constitute an overwhelming threat to privacy and autonomy?

- Do the clinical benefits of nano-enabled medicine outweigh the clinical risks or harms?
- Will the benefits of nanomedicine be equitably and fairly distributed?
- Who has a greater claim to scarce research dollars—those in need of preventive medicine or those in need of curative medicine?
- When weighed against the cost and benefit of other social goods, are we morally obligated to develop and provide the services of personalized medicine to everyone?[2]

9.5   A Special Case of Respect for Communities

Wherever the art of medicine is loved, there is also a love of humanity.

—**Hippocrates of Iphicrates**[37]

The concept of informed consent is deeply embedded in the medical field, requiring that we respect and protect the autonomy and self-governance of the individual human person. The defining characteristic of our moral agency is our ability to weigh the benefits and harms of an action against our own conscience, deeply held values, desires, and future life plan. To deprive a person of the information needed to make critical decisions about one's health and future life is, in some ways, to treat that individual as less than a person. A similar argument can be made for the loss of autonomy related to collecting and distributing intimate medical information, e.g., genetic testing.

The principle of respect for communities takes this concept up a notch to the level of the larger community, obligating us to act in ways that respect the ability of communities to act as autonomous, self-governing agents (Bennett-Woods 2007b; Bennett-Woods 2006; Bennett-Woods and Fisher 2004).[38,39,40] Just as this broadens the scope of who is making the decision, it broadens the scope of whose interests are at stake. Communities have obligations to all current members, as well as to future members, including our children and grandchildren. Concepts of paternalism and intergenerational justice both require that we consider the best interests of those who are unable to consent for themselves. At least two significant issues are raised by nanotechnologies in medicine that have a component of community consent at stake. These include clinical risk along with matters of prioritization, cost, and access.

9.5.1   Clinical Risk

When performing a cost/benefit analysis in medical situations, it is often argued that our primary duty is to avoid needless harm (nonmaleficence). Medical care is rarely without some level of risk. In fact, we often harm patients in the general sense that many medical procedures and treatments are painful or otherwise uncomfortable. Therefore, nonmaleficence applies specifically to avoidable harms related to negligence, carelessness, malpractice, or a failure to properly weigh the benefits against the likely risks and harms. In order to meet the requirements of utility, any harm that is caused must be proportionate to the likely benefit.

It is clear that nanotechnology may offer the possibility of safer and more effective approaches to treatment for a wide array of medical conditions. For example, the ability to more precisely target the delivery of highly toxic chemotherapy drugs will limit the destruction of neighboring tissues. In turn, that will minimize the negative side effects that occur due to the system damage caused by chemotherapy's effects on otherwise healthy tissue and body systems.

On the other hand, the potential risks are hard to predict. The primary clinical risks and uncertainties with nano-engineered drugs, tissues, implants, and other medical interventions include the potential for nanotoxicity of the materials. Toxic responses can occur at the time of implant or as materials degrade within the body. How nanomaterials will interact over time within the body is unknown, but there are reasons to be concerned (Clift 2006; Sweet and Strohm 2006; Colvin 2003).[41,42,43] A similar concern lies with controlling nanomaterials, especially self-directed or self-assembling materials, once they are in the body.[25] As with other applications of nanomaterials, the risks are highly speculative at this time, so the ability to weigh the risks and benefits that a person might when considering more established therapies is limited.

Looking beyond the individual patient, a further risk involves the potential for genetic alterations, particularly at the germline level, to lead to unintended, permanent changes in the human genome that would affect future generations. If the commercial development of therapies gets too far ahead of our understanding of the possible genetic affects of nano-enabled therapies over time, we may inadvertently create problems of a greater magnitude than those with which we started. Similarly, there are concerns about the release of nanoparticles and other materials into the environment through disposal or human excretion. The presence of low concentrations of pharmaceuticals in the environment has been monitored for nearly two decades. There is a concern that environmental exposures to drugs can be harmful to both humans and other organisms (Williams and Cook 2007).[44] The uncertainties about the level of bioactivity and persistence of nano-engineered particles in the environment may warrant a higher level of caution.

An intriguing area of clinical risk is actually social in origin and involves the impact of new technologies on informed consent, and on patient perceptions of illness and health-related behaviors. Research and development in various areas of nanomedicine proceeds with the apparent endorsement of a relatively uninformed public. The predisposition to see the goals and ends of medicine in a positive light makes it relatively easy to minimize the potential concerns. Furthermore, in bioethics, it is recognized that injury and illness are inherently coercive conditions. They threaten autonomy because they render individuals more likely to consent to interventions due to fear, pain, or the perceived lack of alternatives. The heavy cost to society of injury and illness is similarly coercive, and society may support the wide-reaching development of nanomedicine as a sort of magic bullet for everything that goes wrong in the human body.

In actuality, while some applications of nanotechnology are clearly possible, many remain highly theoretical and may prove impractical, overly expensive, inefficient, or only marginally effective in practice. Other applications may prove highly effective, creating a broad demand for their availability. One possible outcome is that the combination of a few specific successes and public overconfidence in medical science may actually increase the overall burden of health care services on society. The stronger the belief that nano-medicine can "fix" anything, the less incentive there is to engage in responsible health practices, such as a healthy diet, exercise, and routine primary and preventive care. The more successful the various therapeutic applications become, the longer the period of time patients with chronic conditions are dependent on the system, especially in cases where medical interventions control without curing. Finally, if applications are pushed into clinical trials or the marketplace too quickly, and outcomes do include untoward effects of toxicity or genetic alteration, the loss of public confidence may undermine support for other applications of nanomedicine that do represent safe and effective developments.

9.5.2 Cost, Prioritization, and Access

Much of this research is being conducted with public funding, and a large segment of health care services in the United States is also covered by publicly funded programs. The percentage of public funding is even higher in most other industrialized countries. Yet, not all segments of society benefit at the same level from new medical technologies. Access to basic health care is a huge issue in the United States, with nearly 47 million people uninsured and many more underinsured. We are making a very clear, moral choice when we decide to invest in medical research rather than the direct provision of needed and currently available services. If the general public was reasonably well informed regarding the cost impact of medical research and new medical technologies, and they understood the extent to which most people have limited access to those technologies, what investment choice would they make? How would they prioritize research and development dollars? We don't know because we never ask them; however, we could pose an interesting and related ethical question. If we invest in these technologies on behalf of the public trust, then are we obligated to also commit to the cost of providing them to everyone?

In the United States, rising health care costs are coupled with rising calls for some form of universal health coverage to every citizen.[2] In a 1983 report entitled *Securing Access to Health Care*, the President's Commission for the Study of Ethical Problems in Medicine and Biomedical and Behavioral Research argued that health care constitutes a social good of "special importance."[45] This importance is based on the relationships among health care and well-being, opportunity, relief from concern, and the interpersonal significance of illness, birth, and death. Although the Commission did not argue specifically for a right to health care, it did conclude that society has an ethical obligation to ensure equitable access to health care for all, balanced by individual obligations, and without excessive burdens on society.[45] What exactly does this mean in an era of rapid advances in high-tech medicine?

Let's say we are successful in developing a personalized medicine that includes molecular surgery, broad spectrum genetic screening and repair, genetically matched medicines, replacement tissues and organs, and other precision diagnostics and therapies. Do these services then become the standard for equitable access for health care? What line would have to be crossed in order for such health care services to be too excessive a burden on society, and who will draw that line? How will individual obligations be established and how will they be enforced? High-tech health care has not typically lowered the cost of health care in the past. Prior experiences don't mean personalized medicine might not result in cost reductions to the system, but there is little precedent on which to base such an expectation. For example, the cost of new technologies such as MRIs and CT scans came down over time; however, their use also increased, as they became the new standard of care. Despite costs eventually coming down for these individual exams, each exam still exceeds the cost of the earlier imaging technologies they replaced.

Expanding the view out a bit further, global health is one of our most serious international challenges. Nanomedicine, like other prior areas of high-tech medical practice, is likely to attract widespread support and use in countries with health care systems capable of delivering them. Consumers in wealthy countries will realize immediate benefits as products become available. On the other hand, the availability of nano-enabled technologies in developing countries will likely be quite a different story.[39]

A panel of international experts agreed on a list of ten biotechnologies that could improve health in the developing world (Daar, Thorsteinsdóttir, Martin, Smith, Nast, and Singer 2002). Nine of the ten on the list involve technologies that have been widely mentioned in conjunction with nanotechnology. They include, in order of importance:

- Molecular diagnostics
- Recombinant vaccine technologies
- Efficient drug and vaccine delivery systems
- Environmental technologies for clean water and bioremediation
- Genetic sequencing of antimicrobial-resistant pathogens
- Drug-targeting bioinformatics
- Nutrient-enhanced crops via genetic modification
- Recombinant technologies for affordable production of therapeutics
- Combinatorial chemistry for drug discovery[46]

Although promising research and development are occurring in all nine of these areas, there is no assurance that successful applications will make their way to those most in need or lead to any measurable short-term benefits for most developing countries.[39] Companies bringing these technologies to market will be interested in generating an appropriate return on investment for shareholders.[39] University and private sector partnerships are increasingly engaging in patenting and restrictive use licensing that limits access to researchers in developing countries or dissemination of knowledge and technology for humanitarian purposes (Meridian Institute 2005).[47] Yet, developing countries have limited resources and limited access to high-tech markets.[39]

9.6   Assessing Options for Action

Nanotechnology in medicine is likely going to be the highest profile nanotechology arena in the public eye. It appears to be moving relatively quickly toward commercialization of early applications, and the public already recognizes and generally supports its potential. Predictions on the time frame of nanotechnology vary wildly; however, a survey of international experts in late 2005 placed most of the nanobiotechnology applications within the next 5-10 years (Hauptman and Sharan 2005).[48] The long-term was defined as beyond 2016 and the very long-term as 2025 or beyond. Only the practical use of nanomachines for therapy and diagnosis inside the body fall into the very long-term category. Even the construction of artificial human organs is placed in the category of before 2020. In societal terms, these are not long time frames, so options for action must take into account the need to move quickly in responding to potential concerns.

Risk assessment will likely take a more precautionary approach as a natural outcome of the myriad of existing regulatory mechanisms and the liability associated with medical applications. However, what is already a long process could become even longer in the face of the uncertainties associated with nano-engineered materials. In addition, the current risk model does not do a very precise job of measuring clinical efficacy and cost effectiveness.[28] Pressures to get products to market, coupled with pressures to reduce the overall cost of health care, suggest the need for a more sophisticated approach to the assessment of medical nanotechnologies, one that gets beneficial applications to market quickly while also maximizing those benefits via the assessment process.

What are less likely to be addressed as ethical considerations are the social risks associated with over-reliance on technology to maintain health and the many issues of cost, prioritization, and access. There is no societal consensus on how to best prioritize research dollars or fund new technologies and adaptations to the larger health care system. For example, there is no attempt to distinguish whether preventative or curative technologies should have priority. Arguments that favor early screening as a means of controlling costs lose their cost efficiency when applied to screening for every possible condition and predisposition. Arguments in favor of acute and chronic care, largely on the basis of obligations to preserve and protect human life, fail to weigh the utilitarian costs of such care against other important social goods. Arguments in favor of curative and regenerative technologies overlook the longer-term societal costs and intergenerational effects of extended life spans. Any reasonable action plan must account for the need to make difficult choices.

Finally, options for action must address the issue of fair and equitable distribution of the benefits of nanomedicine. If left entirely to the market, benefits will continue to flow disproportionately to those members of society with economic means, exacerbating other economic and social disparities that are incidental to poor health. Options for action to address these basic questions of priority, cost, and access all involve tough choices that current decision-making processes fail to make. The choices that are made go well beyond the immediate medical goals. Questions of how best to value and protect human life, maximize the use of scarce societal resources, and strive for a more just distribution of societal opportunities and benefits are profound. They require the informed participation and consent of the larger society. Therefore, the best options for action will involve new approaches to public engagement in both science and technology policy and health care policy.

9.7 Finding Common Ground

Other than generalized concerns about nanotoxicity and genetic implications, there appears to be little opposition to nano-enabled medical advances aimed narrowly at the diagnosis, treatment, or prevention of disease and injury. Most references to the ethical implications of nanotechnology in medicine jump to human enhancement and life extension as the "real" ethical concerns; however, this futuristic focus of concerns may overlook key elements of the larger picture of societal impacts. The rapid advancement of modern medicine and medical technology has given rise to numerous societal impacts. In industrialized countries, whole new categories of ethical issues have popped up in bioethics in the past couple of decades. In the United States, these include prenatal genetic screening and abortion; resuscitation of extremely premature and severely compromised babies (many the direct result of technologies in reproductive medicine, such as in vitro fertilization); advanced life support; people living longer with serious chronic medical conditions resulting in the frail elderly representing our fastest growing demographic; rapidly escalating health care costs; and a growing lack of access to basic, primary health care by millions of people in the midst of the most technologically advanced health care system in the world. At the same time, global disparities in health care increasingly contribute to political and economic instability, the threat of emerging diseases and global pandemics, and tremendous human suffering.

Nanotechnology has the potential to add quality and functionality to the lives of people facing disease and disability. Proceeding with its development is easily justified in terms of utility, nonmaleficence, and beneficence. A healthy population is happier, more productive, and more sustainable, as individuals are able to retain functional independence and achieve their personal potentials. We also have ethical obligations to minimize suffering when we can, and act in ways that serve the welfare of patients and the interest of society.

On the other hand, pursuing nano-enabled medical advances without attention to the larger realities of health care seems short-sighted at best. The evolving conception of health care services as a basic moral right does not currently place much of any responsibility for personal health on the individual. Utility would argue against continuing to provide significant investments in medical technologies that enable citizens to disregard simple lifestyle choices in the maintenance of personal health. At the same time, deep-seated social and economic disparities limit the health choices of many persons, ranging from choices about what to eat, to when and where to seek care if one does become ill or injured. Duties of fidelity to the larger society argue for a balance between personal responsibility for what is owed to society and what can be owed each individual by society. Virtue argues for prudence in intentionally limiting the use of scare resources by individuals as a matter of community well-being, while also showing loyalty and compassion for those in need and those who are most vulnerable among us.

Finally, there are already significant disparities in access to care within the United States and other countries. The disparities loom even larger when compared between nations on a global level. In the face of such disparity in access to the basic components of health, we are also being forced to recognize that our well-being as a society is inextricably tied to the well-being of the global community. Health care and medicine are primary factors in both individual and societal well-being. The fact that current science and technology and policy lacks thoughtful consideration of such impacts is compounded by the unfortunate reality that current health care policy also lacks a systemic focus for dealing with the pace and nature of innovation in the face of deep-seated and complex effects.

As with other issues addressed in this book, if we are to make the most of nanotechnology's potential in medicine, we have to think beyond the narrow technical confines of what is physically possible to what is socially valuable relative to the immediate and long-term needs of the community. In this case, common ground requires a genuine societal dialogue about the limits of health care as a scarce resource and a primary social good.

9.8 Pragmatic Considerations

The impact of NT on the structure and focus of health care will depend on how research and development is targeted, where the breakthroughs occur first, and how the market responds.[2] The timelines for nanomedicine are particularly vague. While some argue there is a wide distance between nanomedicine in concept, published research, and actual clinical application,[29] others suggest that most nanobiotechnologies will be realized within less than the space of a generation.[48] In either case, nanomedicine is not likely to suddenly appear on the horizon, but will roll out unevenly, in fits and starts. Missteps along the way, with respect to either outmoded regulatory systems that delay advances or not enough caution in bringing applications to the market, can result in harm.

More importantly, cost and access issues obligate society to deal realistically with the possibility that nano-enabled advances in biotechnology may add to the overall cost of health care. More broadly, in the absence of a systemic strategy, nanomedicine may also contribute to serious social, economic, and political effects of global disparities in health and trends that accompany an aging population.

Finally, although we did not address it directly in this chapter, another pragmatic reality is that advances in nanomedicine are likely to find their way beyond the treatment of existing disease and dysfunction to the goals of human enhancement and life extension. It is toward this last issue we turn our attention in the next chapter.

9.9 Questions for Thought

1. Do we have a right to health care? If so, are we obligated to provide all medical advances to all patients? If personalized medicine is likely to result in even greater disparities within current health care systems and outcomes across the globe, do we have an ethical obligation to limit its pursuit? What ethical arguments and counterarguments justify your positions?

2. Given the rising costs of health care, particularly in the United States, and the assumption that health care is a finite resource in all countries, how should we prioritize the investment in research and development? To which technologies would you give funding and why? Provide practical and ethical justifications.

PART THREE

The Framework Applied

We thought, because we had power, we had wisdom.

—Stephen Vincent Benét, American author and poet (1898-1943)

10

Case Presentation: NBIC and Human Enhancement

10.1 Case Presentation

> The most difficult part of attaining perfection is finding something to do for an encore.
>
> —Author Unknown[1]

The year is 2050 and you are observing a group of what appear to be young adults interacting in a public space. You are accompanied by a human design specialist who points out the specifications and current functionalities of the beings you are observing. What becomes immediately apparent is that they are all very well-formed in terms of body size, weight, and muscula-ture as well as being particularly attractive. This is due, in part, to the care-ful genetic sorting and manipulation that ensured they would be free from known genetically-based defects and weaknesses, as well as the introduc-tion of various genetic enhancements in the form of preferred traits and abil-ities related to bodily appearance and function. What is not obvious are the genetic enhancements to their immune systems, cognitive development, and metabolism that, in some cases, are the result of particularly useful genes borrowed from other species.

Although they all appear to be the same age, it is impossible to estimate their age very precisely by simply looking at them. The combination of care-ful prenatal genetic selection and ongoing cellular therapies that repair genetic mutations, reverse cellular damage, and maintain optimal cellular functioning has more or less arrested the normal, genetic, and environmen-tally determined aging process. You are assured their life span could easily extend to 150 years.

The primary threat to their health is physical trauma. Your companion points out that she, herself, has an artificial kidney and liver due to an unfortunate accident a few years ago. The organs are a perfect biocompat-ible match to her body, constructed at the molecular level from composite materials and powered by the same basic metabolic reactions as the original organs; however, they are not as susceptible to normal wear, infection, or other disease, so she does not anticipate having to replace them for many years—if ever.

In order to address some of the dangers of ongoing problems with environmental toxins and air pollution, the bodies of the beings you are watching have replaced most of their red blood cells with nano-engineered respirocytes and a web of O_2/CO_2 exchange ports, embedded in their skin, that provide much more effective environmental monitoring and filtering of airborne organisms, toxins, and damaging particulates than the naturally functioning human lung and respiratory system. Other engineered blood cells filter toxins and unwanted microorganisms directly from the bloodstream. With the exception of certain bio/chemical weapons and rare microorganisms that have adapted to avoid detection, the threat of infectious disease has largely been eliminated.

You observe that the mood of those around you is consistently energetic and optimistic. Your companion explains that, in addition to the management of mood, everyone is now capable of extended periods of highly focused concentration thanks to personalized pharmaceutical production units implanted within the body that constantly assess and maintain ideal levels of neurotransmitters and other chemical mediators of optimal neurofunction. Mental disorders resulting from biochemical imbalances in the brain have also been largely eradicated. When you notice people who appear to be communicating but not speaking, she explains that genetically enhanced cognitive abilities, coupled with biocompatible nanoscale circuitry, allow them to communicate telepathically. They can also connect and search virtually every available data repository and perform complex calculations and analyses at roughly the same rate as an early 21st century computer.[2]

The prior description is a composite of projected capabilities that currently appear in the enhancement literature, all of which are on the proverbial drawing board.[3] Many are expected to hit the open market in some form as early as in the next 5-10 years as pharmaceutical companies, medical device manufacturers, and leading research centers build on a rapidly expanding body of knowledge and technological tool set. The question raised by these relatively rapid advancements is whether or not there is a crucial tipping point along the continuum between the YOU of the present and the THEM of a few decades into the future. Is there a point at which the experience of being human could differ so radically from the current human experience that existing models of human biological, psychological, sociological, and cultural functioning will simply no longer apply in a meaningful way? And, if so, should we be worried, hopeful, excited, or terrified?[2]

10.2   The Context Described: Evolution of the Human Person

The most exciting breakthroughs of the 21st century will not occur
because of technology but because of an expanding concept of what it
means to be human.

—John Naisbitt[4]

The traditional goal of medical therapy has been to simply treat an existing
disease or injury, as opposed to treatment for purposes of enhancements
that are intended to go beyond the boundaries of normal human functioning
and health. However, emerging technologies are blurring the line between
the two (Ebbesen and Jensen 2006).[5] For all practical purposes, nearly any
technology that can restore function will likely be able to enhance function
as well. Commonly referred to as NBIC (nanotechnology, biotechnology,
information technology, and cognitive science), the convergence of these
technological capacities provides the basis for manipulating pretty much
every aspect of human functioning (Roco 2003).[6] With the prospect of such
powerful technologies at hand, it seems only logical and inevitable that the
enhancement of basic human capabilities and life extension would follow at
some point.

Is there a practical or ethical difference between injecting nano-enhanced
red blood cells and placing a patient on oxygen, or a respirator, or providing a medication that enhances oxygen intake? Why replace red blood cells
with an exact replica when it might be possible to create a technologically
superior artificial cell capable of carrying enough oxygen to allow you to
survive an extended loss of cardiac or respiratory function? Likewise, if
medical advances (such as human organ transplants) are acceptable, why
shouldn't we develop nano-engineered organs to overcome the scarcity of
human organs?

If it is possible to create neural implants that preserve memory in a brain
damaged by Alzheimer's disease, why not also provide enhanced capacity for general information storage and computational speed? Why stop at
repairing a gene when you can select for, or engineer, genetic enhancements
prior to birth?[7]

The recent history of medicine gives some indication of how readily such
enhancements are likely to be embraced. Assisted reproduction, cardio-pul-
monary resuscitation, and advanced life support have all been considered to
have crossed a moral line when first introduced, yet all have become standards of care in the years since. Despite persistent objections of some faith
and cultural traditions, thousands of babies are born each year by in-vitro
fertilization techniques. We maintain the lives of severely premature neo-
nates and dying elders long past the point at which they would be physi-
ologically capable of functioning on their own. And, although we do have

some qualms about these abilities, the technologies themselves are generally portrayed in the media as examples of the "miracles" of modern medicine.

Human beings have an innate appreciation of, and drive for, perfection. In few places is this more evident than the extraordinary increase in elective cosmetic surgeries or the rising use of genetic screening to detect abnormalities and terminate pregnancies on that basis. Writers such as James Hughes (2004),[8] Ramez Naam (2005),[9] and Ronald Bailey (2005)[10] strongly advocate human performance enhancement as a means of liberating us from the narrow evolutionary boundaries set by genetics.[7] Others, such as Francis Fukuyama (2002)[11] and Bill McKibben (2003),[12] are critical of pursuing these powerful technologies too quickly or at all, envisioning a widening gulf of social disparities or the loss of human identity and values.[7] The popular culture resides somewhere in between, as we find ourselves torn between the popular media images of Frankenstein on one hand and mutant superheroes on the other. Our ongoing fascination with human enhancement seems related to the same general fascination we have with horror films. They entertain precisely because of their power to frighten and horrify us.[7] At the same time, they stimulate our natural curiosity and imagination, while also satisfying our desire for novelty and adventure.

Encouraged in large part by the ELSI initiative of the Human Genome Project, much scholarly attention has been paid to the possibility of human enhancement and radical life extension. For example, while not focused on nanotechnology per se, *Beyond Therapy: Biotechnology and the Pursuit of Happiness*, produced by the President's Council on Bioethics in 2003, addressed the issue of human performance enhancement broadly. The report strikes a cautionary note on proceeding too quickly with biotechnologies intended to enhance human performance and extend the human life span. It warns against a "dangerous utopianism" that neglects the nature and limits of human happiness.[13]

Human performance enhancement is generally at least mentioned in passing in virtually all of the major reports from the NNI, NSF, and others that include medical applications,[14] as well as in counterpart reports such as from the European Technology Platform on NanoMedicine.[15] For example, an NSF/DOC-sponsored report from 2002 strikes a far more optimistic chord than the President's Commission by stating that the long-term implications of human performance enhancement include improvements in:

1. Societal productivity, in terms of well-being as well as economic growth
2. Security from natural and human-generated disasters
3. Individual and group performance and communication
4. Life-long learning, graceful aging, and a healthy life
5. Coherent technological developments and their integration with human activities
6. Human evolution, including individual and cultural evolution[14]

The overall tone of contributors to these various NIH and NSF-sponsored reports is relentlessly positive in the area of human performance enhancement To date, there are no widely embraced guidelines or standards for research and development that is either intended to result in human enhancement capabilities or likely to do so incidentally.

The highly visible, morally controversial, and conflicted depictions of these particular applications of nanotechnologies results in a difficult context to describe. There are little or no facts and figures on which to base an analysis, no regulatory precedents, no good historical analogies, and no societal consensus. In fact, it is likely there is not much genuine social awareness of the issues at hand. Advocates of nanotechnology tend to downplay concerns about radical human enhancement as either too far out in time to concern us now, or simply more fiction than science. Critics of nanotechnology focus on the same concerns in a way that often appears intent upon overshadowing the broader societal dialogue. Neither approach is wise, nor will either stimulate the thoughtful dialogue that needs to happen if we are to make informed choices about technologies as they are developed. We are treading new ground as we re-imagine the foundational assumptions of medicine, health and, ultimately, human identity and meaning.

10.3   Clarifying Purpose

Examining the ethical issues raised by nanotechnology and human performance enhancement is fundamentally an exercise in imagining the whole of the future human enterprise. We are a technological society that places great value on science, innovation, and progress. We are proud of our technological accomplishments and share a well-established faith in the ability of technology to solve problems and enhance our lives. Nowhere is this truer or better deserved than in medicine, but such beliefs are not without challenge. Nanotechnology raises legitimate concerns and novel challenges as well as novel opportunities.

The first challenge is to ask what goals we are attempting to achieve in our search for scientific knowledge and its resulting technologies. Are we obligated to use knowledge once it has been acquired, and for what ends? How and why should we approach this unprecedented opportunity to manipulate life, death, and our experience of everything in between?

The second challenge becomes one of whether or not to set limits on these technologies and, if so, what limits to set and how to enforce them. Does our human inclination toward discovery and manipulation of our environment make it natural, inevitable, and desirable that we use all knowledge and technology? How can we prevent our scientific and technological capabilities from simply outpacing our ability to assess and respond to the deep impact on human culture and society? If we set the limits too tightly, might we fail to realize a new and beneficial level of human achievement, self-actualization, and transcendence? On the other hand, if we fail to set limits will we initiate a form of technological determinism in which too much power is too easily used without the full appreciation of how to use it well?[7] What criteria can be used to assess the benefits and harms for a state of existence for which we have no precedent?

The third challenge involves the extent to which we can, and should, prepare for changes to the definition of human health, the goals of medicine, and perhaps even the definition of a human person.[7] How do we define or redefine health in the face of radical human enhancement technologies? Should genetic, sensory, mechanical, and cognitive enhancements be valued differently? In other words, are some enhancements fundamentally more or less acceptable than others?[7] How can we mediate deep-seated cultural assumptions about social class, political power, market-driven economics, and traditional values of fairness and due process when responding to these inherently disruptive technologies? For example, how can we avoid a tyranny of the enhanced if past forms of intolerance and inequity based on race, gender, age, and culture are simply transferred over to those for whom enhancements are not available?[7]

10.4   Framing the Ethical Questions

The President's Council report cited earlier identifies the same basic ethical concerns discussed regarding medicine in the prior chapter of this book. They include issues of safety and bodily harm (health), unfairness, equality of access, and liberty (freedom and coercion).[13] These lead to ethical questions similar to those posed about nanomedicine; however, they are also different insofar as we have crossed the line from compassionate therapy and restoration of normal functioning to fundamental changes in human abilities and the unprecedented capacity for biological self-design.

1. Should persons be able to pursue human enhancement on the basis of personal autonomy and liberty? Will the compulsory nature of new technologies limit autonomous choice?

2. Will persons who choose not to pursue human enhancement become a vulnerable population in need of protection?

3. Do the utilitarian benefits of human enhancement outweigh the costs or harms?

4. Once available, do we have a duty of fidelity to provide enhancement technologies to all members of the community?

5. If not, how do we minimize the social, economic, and political inequities likely to accrue to those who do not have access?

In addition to these questions, the prospect of human enhancement and life extension also poses deeper and more complex questions.

1. What is the essence of a human person?

2. How much of a human person can be manipulated, replaced, or enhanced beyond its natural function before it becomes something more or less than human?

3. Will the application of performance enhancing technologies before birth and in early childhood be inherently oppressive and a violation of individual human identity?

4. Will human performance enhancement alter the purpose and meaning of suffering, striving, and achievement?

10.5   The Principle of Respect for Communities Applied

The principle of respect for communities requires that we act in ways that respect the ability of communities to act as autonomous, self-governing agents. However, human enhancement may alter key elements of our self-identity and, by extension, our communal identity. It may change the nature of our relationships with each other and the world. Any consideration of communal autonomy must also start with individual autonomy.

10.5.1   The Human Person

At the core of the discussion is the definition of the *human person*, an overtly complex concept whose understanding has been severely fragmented by differing perspectives. The biological sciences view the human person in narrowly biological terms and boundaries. The social sciences have a somewhat expanded view of the person that incorporates self-awareness and the social nature of persons. Religion and philosophy assert yet a deeper meaning and purpose to the concept of a human person. Among these various perspectives, the boundaries of the human person are already highly disputed. As evidenced by ongoing debate on topics such as abortion, stem cell research, and end of life care, we do not agree on when the moral status of a person starts and when it ends. We also don't agree on what the limits of our obligations are to a person; however, these ongoing cultural debates regarding the moral and legal status of human embryos and other entities at the margins of human functioning will ultimately frame the questions and responses needed to evaluate the meaning of human agency and identity in the face of radical life extension and performance enhancements.[2]

What makes a human being human? What makes a human person a person? Many would argue these two concepts are not the same thing, with the human being narrowly defined by gross biological and physiological characteristics such as human DNA. The concept of a human person is often argued to incorporate higher order capacities, such as consciousness, reasoning, self-motivated activity, communication, and self-concept (Warren 1973),[16] or at least the potential for these capacities. In the case of issues such as abortion or treatment for patients in permanent vegetative states, the argument is whether there is a point at which such capacities exist at such a low level as to disqualify an entity from being a person in the moral sense. However, what about the possibility of capacities of consciousness, reasoning, or communication that go well beyond what any current human person holds. Human cognitive enhancement may introduce entirely new capacities for creative thought or reasoning we can't really imagine on this side of it. This raises the intriguing question of whether there are aspects of human functioning, perhaps general intelligence, memory, or emotional capacity, that are so central to human identity that they should not be subject to alteration or manipulation. What percentage of the human body can be replaced or enhanced with artificial or bioengineered components before the entity in question is something more or less than human?

Ultimately, despite the efforts of philosophers, bioethicists, and theologians to construct complex definitions and rationales to answer these questions, the average citizen is somewhat more likely to draw on popular culture and personal experience. The audience was most relieved when the Star Trek character of Data, the android, was saved from being disassembled by his "Creator" thanks to the aggressive defense of Captain Picard on his assertion that rational beings have a right to autonomous choice.[17] Throughout the series, fans cheered on his emerging "personhood" as he struggled to understand and experience human emotion. Perhaps someday, the pioneers of human enhancement technologies will be in the position of cheering on the rest of humanity as we adopt the new technologies in order to "catch-up" and realize our newly imagined human potential.

The question of boundaries is already quite fuzzy. We are all acquainted with people who have cochlear implants, pacemakers, or artificial limbs. Our experience is that they are fundamentally still the same persons, and would likely remain so even if their technology performed a bit above the human norm. The widespread acceptance of cell phones and other wireless technology has more than paved the way for the simple convenience of miniaturized, implantable communication devices that allow for hands-free operation and cannot be easily lost or stolen. Once widely available in the marketplace, can other forms of neuroenhancement be far behind? To what extent will distinguishing traits and abilities become commodities to be purchased in place of gifts to be developed, and how will this affect the symbolic notion of individual human identity?[2]

10.5.2 The Human Person in Community

Bringing us back to the notion of communal autonomy, what, if anything, might constitute the line in the sand that we collectively are unwilling to cross in our definition of persons among us? Perhaps even more importantly, how exactly will we come to some manner of consensus on these questions? The principle of respect for communities rests on the shared values and ongoing dialogue of members of a community. It assumes that communities have a moral right to be self-determining, to make autonomous choices about their own best interests. However, autonomous choices require information and the ability to weigh the likely consequences of our choices. How can we weigh a current existence against one for which we may have little practical insight? How are communities likely to react to an altered conception of human identity? Will human performance enhancement make communities more or less homogenous? Naam suggests that we won't all choose the same enhancements, but will continue to express our individuality through the technological enhancements and abilities we choose to adopt.[9] Will the choice to literally evolve in different directions ultimately build community? Perhaps it will simply fragment and cluster along different lines than it is currently. At the other end of possibilities, an aggressive focus on the evolution of individuals, according to their own preferences, may render the concept of community, as we know it, obsolete. New models of society may emerge that allow us to share space and resources without socially constructed boundaries that currently define where one community starts and another ends. This potential impact on how we live in or out of community may be the single greatest challenge we face in the future.

10.6 The Principle of the Common Good Applied

The principle of the common good calls us to act in ways that respect shared values and promote the common good of communities. The common good is defined, in part, by our collective consensus on the conditions under which human beings thrive. These conditions include the social goods we value, the goals toward which we strive, and the expectations associated with our communal roles. Proponents of human performance enhancement see tremendous opportunity to overcome current problems and bring us closer to the ideals of the common good. Roco and Bainbridge propose a passionate defense of the pursuit of human enhancement.

> At this unique moment in the history of technical achievement, improvement of human performance becomes possible. Caught in the grip of social, political, and economic conflicts, the world hovers between optimism and pessimism. NBIC convergence can give us the means to deal successfully with these challenges by substantially enhancing human mental, physical, and social abilities.[19]

They go on to conclude,

> The twenty-first century could end in world peace, universal prosperity, and evolution to a higher level of compassion and accomplishment. It is hard to find the right metaphor to see a century into the future, but it may be that humanity would become like a single, distributed and interconnected "brain" based in new core pathways of society. This will be an enhancement to the productivity and independence of individuals, giving them greater opportunities to achieve personal goals.[19]

Of course, one man's future utopia is another man's dystopia. Returning to Star Trek for comparison, what these authors describe is uncomfortably close to the Borg, a cybernetically enhanced race that uses their collective intelligence to wander the galaxy assimilating other cultures and destroying those that resist.[18] Humans have not always used their power wisely, peacefully, or with compassion. Serving a common good is not just a matter of consequences, but also a matter of intention. What might the intended goals be of human enhancement and what social goods would we expect to realize in our pursuit of a common good?

10.6.1   Health as an Expanded Social Good

The concept of health has long been deemed a societal good associated with a high level of normal human functioning. According to the World Health Organization, *health* can be defined as "a state of complete physical, mental, and social well-being and not merely the absence of disease or infirmity" (WHO 1948).[20] However, human performance enhancement promises a state of existence that may far exceed a high level of normal functioning. Does the enhanced human simply become the new normal in light of a radically expanded biological potential?

How will we come to define society's best interests in light of potential human enhancement and life extension? The Preamble of the World Health Organization Constitution also specifies that "the enjoyment of the highest attainable standard of health is one of the fundamental rights of every human being without distinction of race, religion, political belief, economic, or social condition."[20] What is the highest attainable standard of health? What interests of society are served if we adopt the assumption that we should provide whatever biotechnical enhancements are available to achieve each individual's personal definition of health and well-being? Perhaps more importantly, what interests of society might we have to sacrifice in order to provide health per this expanded definition.

10.6.2   Other Communal Goals and Social Goods

Of course, society could reject the cost burden of providing enhancement to all members while still leaving the door open for those who wish to pursue it on the basis of personal liberty, another highly valued societal good. Other pressing societal goals such as developing an educated citizenry and providing other basic services may compete to limit support for the routine provision of enhancement technologies. However, this denial of communal responsibility does not necessarily constitute a rejection of the technologies themselves. It simply places them within the scope of the open market, yet another valued social good.

The current fashion of seeking a wide range of cosmetic surgeries solely for the purpose of perceived attractiveness is supported by a general agreement that consenting adults with the ability to pay should be able to seek such services as a matter of personal autonomy and their individual definition of what constitutes optimal physical, mental, and social well-being. This same vein of thought could just as easily be used to justify the use of pharmaceuticals engineered to manipulate mood, enhance cognitive abilities, or increase physical endurance. On the other hand, our liberty interests may also be compromised by the open market availability of enhancement technologies. If my ability to acquire and hold a job depends on how well I compete with my cognitively enhanced colleagues, then the technology takes on a compulsory character than may well undermine personal liberty and other freedoms we hold in high esteem.

10.6.3 Societal Roles

Modern society depends on a complex web of individuals filling various roles within the society, and working together to achieve the common good. Let's consider the role of the physician (as well as other health care providers), commonly associated with the societal goods of medicine and health. As the applied definition of health expands, then so does the acceptability of, and demand for, enhancement technologies. In fact, at the point at which enhancement technologies begin to give individuals an edge in the job market or other social spheres, it will become harder to claim that such services do not fall into the category of basic health care. Current conceptions of elective treatment may be subject to revision by societal consensus and a demand to include enhancement as a routine form of primary care.

In medicine, the *standard of care* is both a legal and clinical term for the actions any prudent physician or other health care provider would take in a given situation. Just as other successful advances in life-saving technology are eventually established as the standard of care in those specific life-threatening situations, radical prevention and enhancement strategies may become the standard of care in primary and acute care settings, reflecting both acceptance by patients and providers of altered expectations in the outcomes of health care encounters.[2]

In response, it is not hard to imagine the traditional role of physician as healer expanding to physician as designer and enhancer. The plethora of current, reality-based television shows that chronicle "extreme makeovers" in the form of radical cosmetic surgeries, gastric bypass, and other inter-ventions, portray physicians as saviors—altering the lives and very identi-ties of their patients. Embedded at the center of life and death struggles, the health professions have always been imbued with a certain heroic qual-ity. However, if human enhancement technologies come to be viewed as compulsory, driven largely by the goals of the marketplace, physicians as designers and enhancers may come to be seen in a much less heroic light. Will these new arenas of practice remain consistent with the altruistic identity of healer, or will they eventually come to be seen as manipulative, oppressive, and exploitive? The potential disintegration of roles underly-ing critical social goods may prove a greater harm than is currently being anticipated.

10.7   The Principle of Social Justice Applied

The principle of social justice requires that we act in ways that maximize the just distribution of benefits and burdens within and among communities. In the same way that the uneven distribution of nanomedicine may exacerbate existing societal disparities, human enhancement could take such dispari-ties further, yet on both a national and global scale. Social justice calls us to look within and beyond our own community borders and consider our inter-ests in light of those of all other communities. This spectrum also includes communities of the future in the form of future generations.

10.7.1   Tyranny of the Enhanced

How will enhancements come to be valued? Will society become stratified on the basis of enhancement levels in much the same way it has traditionally been stratified on the basis of race, gender, or class? Will the enhanced come to see themselves as entitled to a dominant social position by virtue of their parents' good up-front design choices (Fukuyama 2002).[11] Critics of enhance-ment generally see it as radically upending any hope of a level playing field for those members of society that currently do not have access to basic needs, let alone advanced technologies that would allow them to compete fairly in the open marketplace.

A particularly intriguing question is raised by the emphasis on this research in the military. The military is currently a primary investor in human enhancement technologies as was noted back in Chapter 7. This fact raises the interesting question of what will happen to "future warriors" coming out of the military. Will they be accepted, and perhaps even sought out, for their interesting spectrum of abilities? Will they be relegated to law enforcement and other jobs that make use of their specific warrior capabili-ties? Will they be feared, subject to an underlying mistrust of their power and the potential for that power to be turned against the larger society. Will they be tempted to use their capabilities toward criminal ends? Will they pose a public threat or simply be treated like they do?

10.7.2 Tyranny of the Elders

What are the implications of a longer life span? Death is a normal, albeit often unwelcome, part of nature. Death, among other things, imbues the human life span with meaning. It is the final milestone in a human life and we live our lives with that inevitability clearly ahead of us. We surround it with ritual and mark it with accomplishments large and small. We also resist it and such resistance meets its epitome in the field of anti-aging medicine, which is directly associated with many of the projected advances in nano-enabled biotechnology.

Proponents of anti-aging medicine generally predict that the human life span will increase to between 120 and 150 years within the next 50 years (Mykytyn 2006).[21] Citing a Freedonia Group Study in 2005, Mykytyn points out that the anti-aging marketplace represents a $20 billion expenditure by consumers. However, the cost of expanded life spans goes well beyond the pocketbook of the individual consumer. It represents a prioritization of research dollars and investment away from other targeted research. It means higher costs, over longer periods of time, of social programs such as Medicare and Social Security.

Perhaps most importantly, it represents a potentially huge societal cost, as people live twice as long, consume twice as many resources, and simply take up "space," both literally and figuratively. In China's mandate for one-child families, we do have a modern precedent of what to do with too many people. Restrictions on reproductive autonomy would seem inevitable, as more and more resources are consumed by fewer and fewer people. The trend toward accumulation of great wealth by a relatively small number individuals may increase as less wealth is distributed across the generations. Retirement before the age of 100 will be unlikely for most people, leaving little room in the job market for nearly two entire generations.

On the positive side, increased life span may be the answer to decreasing fertility rates across the globe. If the population is aging anyway, wouldn't it make sense to help them age gracefully and productively? Elders living longer could be an economic engine and a fount of valued wisdom and stability as we rapidly work our way into the future.

10.8 Assessing Options for Action

Too much is at stake to believe that human performance enhancement can simply be stopped in its tracks. The line between therapy and enhancement is too fine and the natural human disposition is to compete effectively. Efforts to ban human enhancement technologies will largely fail, driving research and development activities underground or into friendlier territory.

Khushf (2003)[22] suggests that our best option is to set aside any current societal consensus on what we think the goals of community and human thriving should be based on what they have been in the past, and "seek to form a new consensus, asking how enhancement should be understood, and what forms such enhancement should and should not take."[22] He contends that the possibility of something "genuinely new" is absent from current dialogues that tend to fall on the extremes of support and non-support for human enhancement technologies. In essence, our only real option may be to stimulate dialogue on a broad scale that is at least as creative and strategic about the societal goals we wish to achieve as scientists, engineers, and entrepreneurs are being about the science itself.

10.9   Finding Common Ground

Did human beings stop developing once they hit the top of the evolutionary ladder, or have we always been in a process of unfolding human potential? The rapid advance and sheer complexity of human society and culture is testament to a certain ongoing level of evolutionary change within the species. In fact, there is a contemporary tendency to refer generally and metaphorically to emerging technologies as leading to the next step in human evolution.[2] Some commentators even predict the emergence of a new species of human beings. Technology advocate Ramez Naam posits this evolutionary process as inevitable, labeling modern humans a "phase shift" in biology.

> We are not the end point of evolution—there is no such thing. We are just an intermediate step on one branch of the tree of life. But from this point on, we can choose the directions in which we can grow and change. We can choose new states that benefit us and benefit our children, rather than benefiting our genes.[9]

The prospect of achieving a higher order of human potential is tremendously appealing in the current face of daunting problems in all realms of human society. Freedom from physical and mental infirmities, coupled with expanded abilities to pursue the activities and interests that inspire and enlighten us, may liberate much needed levels of human energy, creativity, and potential. Assuming that an enhanced existence will also lead to a more enlightened perspective opens the door to utopian views of a future in which the highest human ideals can finally be realized on a broad scale. In evolutionary terms, our survival may depend on our ability to adapt to a higher level of functioning in a new and more demanding environment. In the face of accelerating pace, increased complexity, and a high order of uncertainty, human performance enhancement may be the key to our continued evolutionary success.

Others have a much dimmer view of efforts to exceed our natural human boundaries, one that is equally compelling. Striving for perfection holds great meaning until the achievement of perfection is a result of technological intervention rather than personal effort. Thinking faster and thinking well is not the same thing. Enhancement implies power and power can be abused. The prospect of a lifetime that lasts 150 years or more may simply result in an extended childhood and adolescence as we try to figure out what to do with all that time. The normal human milestones that mark the progression of a human life may become skewed. The pressure to make a difference and to leave a legacy may kick in much later or burn out much sooner. Wonder, curiosity, joy, sorrow, and all the other emotions that shape human character may dissipate into a technology-induced absence of human affect and, ultimately, human meaning.

As with so many other issues raised in this book, reality is not likely to reside primarily at one end or the other. There is much precedent for the abuse of power by dominant forces in the social order. Common ground will require assurances and safeguards that can forestall a tyranny of the enhanced, particularly during the early stages of availability when access is likely to be limited to an already privileged few.

There is also reason to be concerned that human nature requires the trials and tribulations of life to stimulate personal growth and to nurture wisdom and insight. On the other hand, technology has long shaped human experience and identity, enabling an extraordinary flourishing over a relatively short time frame in the scope of human evolution. It will continue to do so in any form. Technologies that enable human enhancement and life extension will represent the same two-edged sword of all previous technologies. Finding common ground will require that we acknowledge both the potential for benefit and for harm.

In order to attend to the various ethical questions raised, common ground must strive for the following:

1. Respect for the essential dignity of the human person, however it comes to be defined

2. Support for the flourishing of all human communities

3. Dynamic, ongoing, representative dialogue that continuously refines the definition of common good

4. Commitment to social justice, including just treatment of future generations

5. Clarity and transparency of intention and goals

6. Ongoing assessment of societal impacts

7. Collective willingness to change paths quickly in the face of untoward or unwelcome outcomes

COMMENTARY

Nanoscale Sciences and Technology and the Framework of Ableism

Gregor Wolbring

Many nano-taxonomies exist which highlight numerous nano-fields, processes, and products. An evaluation of nano as a whole is difficult, if not impossible. Every nano-field, process, and product has to be evaluated differently as each poses distinct challenges and rewards, and each has a unique dynamic, around which a positive could turn negative and a negative could be eliminated and turned to a positive. Nano-evaluations become even more multifaceted if one takes into account that different nano-fields, processes, and products allow for the interaction and convergence with other technologies and sciences.[1] It also does not help that the discourses around the development and application of ethics theories, which are supposed to allow for a "positive, good" governance of science and technology, are intrinsically biased.[2]

One of the most consequential outcomes of nanoscale science might be the generation of products, processes, and knowledge that allow for "improvement" and modification of the human and other biological bodies (structure, function, capabilities) beyond their species-typical boundaries.[3] Many arguments exist for and against enhancement[3] and the validity of any given argument depends on cultural, economic, ethical, moral, spiritual, and political frameworks, the discourse between social groups, and the judgment of and by social groups. Two concepts on which the validity of many arguments hinge are the concepts of ableism[4,5] and transhumanism.[6]

Ableism is a set of beliefs, processes, and practices that produce a particular kind of understanding of one's self, one's body, and one's relationship with others of one's species, other species, and one's environment. Ableism exhibits the favoritism of certain abilities that are projected as essential, while at the same time labeling deviation (real or perceived) from these essential abilities as a diminished state leading to the justification of a variety of other "isms."[4,5]

Ableism reflects the sentiment of certain social groups and social structures to cherish and promote certain abilities, such as productivity, competitiveness, and consuming over others such as empathy, compassion, and kindness.[4,5] Ableism and favoritism of certain abilities shaped, and continues to shape, numerous areas, such as human security,[7] social cohesion,[8] many social policies, and relationships among social groups, between individuals and countries, between humans and non-humans, and between humans and their environment.

Ableism is employed against the traditional as impaired, as subnormative perceived disabled people, and against the "traditional as normative species typical perceived non-disabled people."[4,5] Ableism is also among others partly driving GDP-ism, consumerism, ageism, caste-ism, ethnicism, speciesism, and other ableisms based on cognitive abilities, antienvironmentalism, and superiorism.[4,5]

Ableism will become more prevalent and severe with the anticipated ability of new and emerging sciences and technologies to:

1. generate human bodily enhancements in many shapes and forms with an accompanying ability divide and the appearance of the external and internal techno-poor disabled

2. generate, modify, and ability enhance non-human life forms

3. separate cognitive functioning from the human body

4. modify humans to deal with the aftermath of antienvironmentalism

5. generate products atom by atom, which moves the trade from nature-based commodities towards atomic-generated commodities and will change the way we trade

We can already observe a changing understanding of one's self, one's body, and one's relationship with others of one's species, other species, and one's environment. New forms of ableism, which are often sold as a solution to the consequences of other ableism-based "ism"s, are appearing.

The second concept, transhumanism, is described in the FAQ of the World Transhumanist Association as follows:[6]

> Transhumanism is a way of thinking about the future that is based on the premise that the human species in its current form does not represent the end of our development, but rather a comparatively early phase. The intellectual and cultural movement that affirms the possibility and desirability of fundamentally improving the human condition through applied reason, especially by developing and making widely available, technologies to eliminate aging and to greatly enhance human intellectual, physical, and psychological capacities...

Transhumanized versions of health, disease, and other concepts are developing.[3] A variety of transhumanized versions of ableism[4,5] (e.g. human, animal, and environmental) are appearing. The human version entails a set of beliefs, processes, and practices that perceive the improvement of the human body and functioning beyond species typical boundaries as essential. The emerging field of enhancement medicine provides the tool to live out this version of ableism.[3] It elevates the existing medicalization dynamic of the human body to its ultimate endpoint, where enhancements beyond species-typical body structures and functioning are seen as a therapeutic intervention (transhumanization of medicalization).[3]

The environmental version entails a set of beliefs, processes, and practices which champions: a) the enhancement of especially the Homo sapiens body beyond species typical boundaries; and b) the shaping of the environment (geo-engineering, gated biospheres, etc.) as the solution for the climate change to come. Finally, the animal version entails a set of beliefs, processes, and practices that champions the cognitive enhancement of animal species beyond species typical boundaries as a remedy for ableism-driven speciesism.[9]

Dealing with ableism and transhumanism is essential if one wants to diminish, reverse, and prevent the strife one can expect in regards to the disruptive potential of many nanoscale science and technology products, such as the enhancement of animals (which will redefine the relationship between humans and animals), immortality and longevity research (which will redefine intergenerational relationships), molecular manufacturing of material from the atom level (which will redefine the trade system as we have today), and products intended to modify the appearance and functioning of the human body beyond existing norms and species-typical boundaries (which will redefine self-identity and how we see other people, other species, the environment, and ourselves). Addressing ableism, through the wide-ranging discipline of ability studies, could help deal with the challenges ahead of us, such as a real and durable sustainable equity and equality for any one country, group, or individual and a real culture of peace.

References

1. Wolbring, G. 2007. "Social and ethical issues of nanotechnologies." *ISO-FOCUS* 4(4):40-42, http://www.bioethicsanddisability.org/isofocus.html

2. Wolbring, G. 2003. "Disability rights approach towards bioethics." *Journal of Disability Studies* 14(3):154-180.

3. Wolbring, G. 2005. *HTA Initiative #23—The triangle of enhancement medicine, disabled people, and the concept of health: A new challenge for HTA, health research, and health policy.* ISBN 1-894927-36-2 (Print); ISBN 1-894927-37-0 (Online), ISSN: 1706-7855. http://www.ihe.ca/documents/hta/HTA-FR23.pdf

4. Wolbring, G. 2007. *NBICS, other convergences, ableism, and the culture of peace.* Innovationwatch.com webpage, http://www.innovationwatch.com/choiceisyours/choiceisyours-2007-04-15.htm

5. Wolbring, G. 2007. *What convergence is in the cards for future scientists?* Conference presentation, Vienna, May 2007. Hosted on International Center for Bioethics Culture and Disability webpage, http://www.bioethicsand-disability.org/convergence

6. World Transhumanist Association. 2003. *The transhumanist FAQ—A general introduction—Version 2.1.* World Transhumanist Association webpage, http://www.transhumanism.org/index.php/WTA/faq21/46/

7. Wolbring, G. 2006. *Human security and NBICS.* Innovationwatch.com webpage, http://www.innovationwatch.com/choiceisyours/choiceisyours-2006-12-30.htm

8. Wolbring, G. 2007. *NBICS and social cohesion.* Innovationwatch.com webpage, http://www.innovationwatch.com/choiceisyours/choiceisyours-2007-01-15.htm

9. Wolbring, G. 2007. *Enhancement of animals.* Innovationwatch.com webpage, http://www.innovationwatch.com/choiceisyours/choiceisyours-2007-03-15.htm

Dr. Gregor Wolbring is a biochemist, bioethicist, scholar in governance of science and technology and ability studies, and health policy researcher at the University of Calgary, Alberta, Canada. He is a part-time professor at the Faculty of Law, University of Ottawa, Canada (Sept. 2007). He is, among others, a member of the Center for Nanotechnology and Society at Arizona State University; a member of the Canadian Advisory Committees for ISO/TC 229, *Nanotechnologies*, as well as on the editorial team for the Nanotechnology for Development portal of the Development Gateway Foundation; Chair of the Bioethics Taskforce of Disabled People's International; and member of the Executive of the Canadian Commission for UNESCO. He publishes the Bioethics, Culture and Disability website (http://www.bioethicsanddisability.org, writes the biweekly column *The Choice is Yours*, http://www.innovationwatch.com/commentary_choiceisyours.htm, and a blog on new and emerging technologies at http//:www.wolbring.wordpress.com.

10.10 Pragmatic Considerations

In many ways, this chapter is intended to be a cautionary tale. In essence, all nano-roads lead to human performance enhancement and life extension. Advancements in nanoscience and nanomaterial development and manufacturing will lay the foundation for all nanotechnology applications including those involved in nanotechnology. For example, applications involving intimate human-machine interfaces may well have their origins in the information technology and telecommunications industries. The military is aggressively pursuing a whole range of enhancement technologies in their pursuit of the Future Force Warrior. As with prior military technologies, these will eventually make their way into medicine and other industries in the form of consumer services and goods that enhance human "survivability" and "superiority" in essentially the same spirit they are intended to enhance soldier survivability and superiority (i.e., lethality and fightability in military jargon) on the battlefield. Along a slightly different vein, continued environmental degradation of air and water, or failure to address emerging environmental threats such as climate change, may render human enhancement a requirement if the species is to survive into the future. Finally, there is no reason to expect that medical advances that restore function won't also be used to enhance it. Sports doping is an obvious example of the use of medical tools for non-medical ends and the difficulty in preventing it from happening. A failure to begin to develop a holistic and systems-oriented vision of science and technology will increasingly allow science and technology to define the human good with increasingly less input from the actual humans.

The complexity and relative messiness of society at large places us at a serious disadvantage when trying to decide how to move forward in the face of rapid pace and uncertainty. Nonetheless, move forward we will. The true pragmatists have little choice but to roll up their collective sleeves and try to find the wisdom to guide speed and direction in which these technologies ultimately alter the human condition. To do so ethically requires, at a minimum, that human enhancement technologies are welcomed or marginalized in order to respect the essence and will of the larger community and the best of our human values and aspirations. The values of liberty, knowledge, compassion, curiosity, and human striving must be carefully weighed against the human tendencies to also act in narrowly self-serving ways and to abuse power once they have it.

In addition, these technologies must serve a common good, requiring that we engage in a careful differentiation of the benefits from the harms, and that we consciously work to minimize the harms that are inevitable. Finally, pragmatism dictates that we seek a form of social justice so that the proposed nano-divide does not further fragment our increasingly interdependent human community.

10.11   Questions for Thought

1. Is there a morally significant difference between repairing and enhancing basic human abilities? Doesn't all technology enhance human performance to some extent? Better nutrition increased our height and average IQ. Sanitation and antibiotics increased our life

span. Why shouldn't we develop technologies that allow us to think faster, see and hear better, be more physically fit, or extend our life span by another 50 years?

2. Propose a system in which human enhancement and life extension technologies could be made available that would meet the requirements of true informed consent (voluntary and competent), minimize the potential harms or abuse of the technologies, and ensure a measure of social justice.

11

The Ethical Agenda for NT

11.1 The Pressing Questions

> We live in a society exquisitely dependent on science and technology, in
> which hardly anyone knows anything about science and technology.
>
> —Carl Sagan

The lack of public understanding of science is well documented.[1] Whether
due to weak K-12 science curricula, over-simplified and sensationalized
media accounts, or simple lack of interest, our ability to assess the goals and
outcomes of science and technology on a broad social level is increasingly
compromised. The effects of rapid pace, increasing complexity, and greater
uncertainty complicate decisions regarding investment, regulatory oversight,
and global economic strategy. At no time has our need for broadly situated
expertise and an informed citizenship been greater. Yet, nanotechnology is
poised to present the most sweeping and significant challenges we have yet
faced with respect to nanotechnology, and we seem ill-prepared to confront
it with the practical wisdom it deserves.

An early warning was issued in a widely read article entitled *Mind the Gap:
Science and Ethics in Nanotechnology* (Mnyusiwalla, Darr, and Singer 2003).[2]
These authors contended that, despite the availability of research funds,
research related to nanotechnology and the environment, ethics, economics, legal, and safety issues, were not being taken seriously or occurring on a
sufficiently large scale to make a difference. They cited a scarcity of publications on the ethics or societal implications of nanotechnology and warned
that the increasing gap between ethics and the science could "derail" nanotechnology somewhere along the line. Based on lessons learned from the
effort to address the societal implications of biotechnology and genomics,
they made a number of suggestions. The first was to fund ethical, legal, and
social implications research in much the same way as the Human Genome
Project.[3] In the United States, the NNI has done so, exceeding the 3-5% set
aside in the Human Genome Project. On the other hand, these authors also
warned against the "navel-gazing" approach of early studies in the Human
Genome Project, recommending instead a larger, interdisciplinary research

platform and an "intersectional approach" that allows ethicists and social scientists to interact regularly with scientists, NGOs, and other activist and public interest groups, government and industry. Although it is probably too early to pass judgment, the emerging infrastructure of the NNI does seem to have attempted a broader and more interdisciplinary approach. Academic centers have been funded at Arizona State University,[4] University of California at Santa Barbara,[5] and the University of South Carolina.[6] In addition, there is a requirement that societal implications research be included in NNI funding proposals. Other recommendations included strengthening the capacity to do such research through funded training, involving developing countries and promoting public engagement in the process.[2] Of these three recommendations, the NNI's efforts to "win the hearts and minds" of the public is the most apparent, with strong emphasis on public education and acceptance (Bond 2004).[7]

On the surface, this all seems to be leading us in a positive direction. However, nagging questions remain.

- Is this a serious effort to resolve difficult moral and pragmatic issues, or window-dressing designed to comfort or deflect rather than confront?
- Are the resources adequate to the task?
- Have the resources been allocated in the most effective manner to those most capable of using them well?
- Will the various attempts to address societal implications respond as quickly as the discoveries and advances are likely to unfold?
- When there are missteps and untoward outcomes, how quickly and effectively will we respond to change course?
- Do we have the collective political, economic, social, cultural, and moral will to make hard choices?

Keiper (2003)[8] suggests that what debate does occur is likely to focus on how best to target our investment and how to mitigate the overt and measurable environmental and health issues. However, he also poses deeper questions.

> Finally, we will be charged with rethinking our place in the universe. Our new powers of precision and perfection could lead us to a deeper appreciation for life—or they could make us lose all respect for the imperfect world we inhabit and the imperfect beings we have always been. The era of nanotechnology may be one of hubris and overreach, where we use our godlike powers to make the world anew. Is there room for wonder in a future where atoms march at our command?[8]

Given the likelihood that nanotechnology is likely to unfold across many industry sectors in a relatively short period of time, it will be easy to focus our attention on the disparate benefits and harms of individual advances. The fragmentary effect of spreading our attention thinly and reactively across such a wide space will make it much easier to lose sight of our higher goals and aspirations. Who will be the keeper of the larger social vision? Who will control the agenda, make the decisions, and implement a direction that best serves society as a whole?

11.2 The Players

The most common way of categorizing the players in the nanotechnology debate is to place them in somewhat distinct categories by social role: scientists, engineers, politicians, corporate interests, the media, consumer advocates, environmental activists, the general public, and so on. However, this approach to grouping tends to overlook the more complex nature of society and its diverse interests. The boundaries between these groups are inherently fuzzy. Scientists and engineers can also represent business interests. Politicians are also de facto members of the media, carefully crafting their messages and manipulating media outlets to represent particular views. Business leaders can be strong advocates for environmental issues. Citizens can, on their own initiative, develop high levels of expertise and take on activist roles that make them more comparable to industry consultants than the general public. Finally, we are all parents or close relatives with a shared interest in the future of the next generation. If we tightly label individual groups, and then ascribe to them a single set of either altruistic or self-serving intentions, we undermine our ability to use our natural diversity as a source of innovation and creativity within the societal dialogue.

Rather than portray the players in the nanotechnology debate in traditional categories, as others have done previously and well (Berube 2006; Schummer 2005),[9,10] I would like to re-define the boundaries a bit more on the basis of the nature of the likely contribution to the larger debate. Each of these groupings is populated by an inherently diverse, non-homogenous constituency that brings a slightly different set of abilities and assumptions to the table, as well as a slightly different base of power.

11.2.1 The Funders

This arena consists of everyone from primary financial investors, including Congress, to those entities and individuals responsible for prioritizing and allocating resources to research and development, to corporate interests targeting growth opportunities, to individual consumers who essentially vote their funding priorities with their own pocketbooks in the marketplace. The core assumption of funders is that resources are limited, and therefore, must be managed. Money is power. The skill set of effective funders includes evaluating the likely return on investment, assessing financial and material risks in both short- and long terms, prioritizing resources, and planning for scarcity. Their primary ethical challenge is utility and maximizing benefits. Their funding/purchasing decisions effectively determine what does or does not get done throughout the entire social order.

11.2.2   The Thinkers

The arena of the thinkers includes academics and scholars from every discipline, engineers and technical experts, policy strategists, visionaries and futurists, and anyone who has a well-established habit of asking questions. The core assumption of these folks is that knowledge is good and worth reflecting upon. Knowledge is power. At their best, thinkers relentlessly question the body of knowledge, collect information and recognize patterns, synthesize multiple viewpoints, systematically test their observations and ideas, reflect deeply about the implications of what they know and do not know, and ask questions of meaning and purpose. Their primary ethical challenges are fidelity to the truth and humility in the face what they do not yet know. Whether following basic scientific method or constructing philosophical debate, thinkers provide the conceptual framework through which we understand our world and each other.

11.2.3   The Communicators

Communicators provide the means by which data become information and are passed along to others. Whether a reporter on the nightly news, a junior high science teacher, a Congressional speech writer, a marketing guru, or the instigator of debates at the family dinner table, communicators are the initiators and vehicles of social engagement and debate. Their core assumption is that what we do is based on what we know as well as the context in which we know and understand it. Communication is power. The skills of the effective communicator include framing a clear and consistent message, sorting extraneous information from the useful information, giving voice to multiple and conflicting viewpoints, generating interest, and creating dialogue where it might not otherwise naturally occur. Their primary ethical challenges are justice and fairness in what they choose to communicate and the manner in which they choose to communicate it. The communicators are the critical arbiters of the tone and trajectory for social dialogue.

11.2.4   The Arenas Combined

As you can see, any implied boundaries are, by design, extremely porous, representing a more flexible and organic delineation of social categories that recognizes the potential for each person to be a "player" in every arena. A player is basically anyone who is meaningfully engaged in one or more arenas and capable of bringing a set of identified skills to the table. The member of Congress may wield more funding power than you do when shopping at the local department store, but the simple reality is that there are a lot more of you than there are members of Congress. The physicist may have an exquisite understanding of quantum mechanics, but the sociologist or ethicist may have important insights into what this might mean as a measure of human identity and control. Well-funded liberal and conservative media outlets may reach millions of homes simultaneously, but a single blogger with the facts can change tomorrow's headlines in a single, well-timed post.

If what is meant by the rather amorphous term "public engagement" is that the public itself, which by definition should logically include a representative sample of everyone, becomes a player in the societal dialogue about nano-technology, then we must broaden both our conception of what the public is and the forums in which public engagement occurs. The potential for a strategically balanced mix of people, broadly representing each arena, could form the conceptual basis of a more coherent and cohesive infrastructure for true societal assessment and response to the challenges and opportunities of nanotechnology. Balanced ethical analysis, by definition, requires this larger and more flexible view of the public sphere.

The primary obstacle to public engagement seems to be resistance to a more egalitarian openness to multiple sources of authority. We are finally begin-ning to recognize the value of interdisciplinarity in science (Jotterand 2006),[11] an inescapable insight of nanotechnology which itself represents the conver-gence of several scientific fields that previously held competing explanatory and methodological models. Calls for the further integration that includes the social sciences and humanities is a logical extension, especially if we are to hold science accountable for its societal impact. An additional challenge to the authority of science lies in the realm of politics and economics. For better or worse, the scientific enterprise finds itself ever more closely tied to politics and economics.[11] Attempts to deny this reality or return to the days of science "for the sake of science" (if there ever was such a time), are not only fruitless, but counterintuitive in the face of the potential economic and politi-cal impact of nanotechnology.

Finally, accepting the public as a legitimate source of authority will be far easier said than done. Consider the evolution of informed consent in medi-cine. The dominant model of the physician/patient relationship for most of the history of medical practice has been paternalism. The physician, by virtue of training and experience, is assumed to be in the best position to make decisions on behalf of the patient. The patient's primary obligation is to follow orders. This model has been severely challenged in recent years and is slowly giving way to a much stronger concept of patient autonomy and partnership. In the process, physicians have had to give up certain ele-ments of control while patients have been forced or empowered (depend-ing on your point of view) to take more responsibility for their health and decisions regarding their health. In further support of their newfound autonomy, patients now have unprecedented access to medical information ranging from technical studies published in obscure medical journals to clinical outcomes data on hospitals and individual physicians published on the internet. Not all patients fully appreciate this newfound power; however, most do. While the shift has been a somewhat painful one, it also been a positive one for the health care enterprise generally.

A similar paradigm shift has to take place if true public engagement in science and technology assessment is to occur. Science itself represents an elite and largely self-defined subculture. Science promotes a particular version of truth that describes the world primarily in terms of derivative theory, hypotheses, and data-driven logic. That which can be observed and measured is valued far more than those elements of reality that tend to defy prediction or measurement. Narrowly constructed concepts of validity and reliability are the basis for defining useful knowledge, along with peer-review, which is itself an inherently political process. Beliefs, attitudes, and assumptions that fail the tests of scientific rigor are inherently suspect and relegated to secondary sources of authority at best. Those who espouse such beliefs, attitudes, and assumptions are often labeled misguided, misinformed, unscientific, non-technical, non-expert, and otherwise incapable of an adequate enough understanding of science to warrant a voice in the process. The general public is tacitly deemed more of a barrier to be breached than an asset to be used in the process. This is readily apparent in the NNI's focus on educating the public on the science of nanotechnology to "inoculate the body politic from the virus of fear."[7] There is little or no suggestion that scientists themselves might benefit from exposure to the voices of other-than-science. Most importantly, there is no comparably well-defined position or role for such voices in the discussion.

However, history has demonstrated the value of the public voice. Social and environmental activists, politicians, regulators, artists, and the media have often been labeled misguided, misinformed, sensationalist, or even ignorant. The well-publicized debate over global warming is an interesting, and perhaps, instructive example of politics and industry using the incessant message that "the science isn't in yet" to justify inaction, despite mounting scientific and anecdotal evidence along with basic common sense. In this case, the debate among the scientists eventually appears to have more or less resolved itself in favor of global warming; however, large segments of the global public were already there due to other-informed and other-guided voices from literature, art, religion, news media, entertainment, and popular culture.

The point is not to suggest that every individual voice be given equal weight, but to argue the need for a reasonable space in which the legitimacy of the various public voices can be examined. Any particular voice, public or otherwise, can be shown to be narrowly self-interested, overtly biased, closed to meaningful dialogue, or overly reliant on hyperbole and poorly supported claims; and while this doesn't render them irrelevant and totally without merit, it does warrant a higher level of scrutiny. Insights from either end of the nanotechnology support/non-support continuum are valuable if they can be substantiated with verifiable information, relevant precedent, or well-reasoned moral and pragmatic claims.

The keys to a true public engagement are inclusion, trust, and transparency. The "underinformed" cannot know what the "experts" are not willing to tell them. Information cannot be dumbed down, oversimplified, given a positive spin, or overtly withheld on the grounds that the audience either can't understand it or will react badly to what they do understand. The public sphere must be trusted with full disclosure, and scientists, politicians, and industry must be willing to give respectful credence to the public's interpretation of that information as it is filtered through their values and expectations. Public engagement, at its very essence, means a collective process of exploring human nature and its attendant questions of human purpose, value, and meaning. It moves us beyond the narrow roles of scientist, politician, and CEO, allowing us to interact as friends and neighbors in pursuit of a common future.

COMMENTARY

Stakeholder Participation in Nanotechnology Policy Debates

David M. Berube

Having recently researched and written about stakeholders and nanotechnology, there appeared to be a disconnect between what is meant by the term and how it was used by different interest groups as well as individuals in the debates over environmental health and safety, and other societal implications of applied nanoscience and nanotechnology. Beyond the obvious conflation of the term shareholder into stakeholder, my complaints remain regarding government legislation to communiqués by public service interest groups, often coded as nongovernmental organizations. While they pledge to have the concerns of all relevant stakeholders at heart, they mostly have not taken steps to define who composes these groups and what can be done to involve them.

This brief piece posits a few controversial hypotheses. First, who are the true stakeholders in the nanotechnology debate? And are all stakeholders equal? Second, are we really interested in broad *stakeholder participation*? If so, what does that mean and what can be done to improve *stakeholder participation* in nanotechnology policy making? Third, are some players using the rubric of *stakeholder participation* as a tactic to advance self-interests maybe even at odds with a meaningful, more altruistic public agenda?

Before answering these questions, it is important to understand the problems I have when using the word *nanotechnology*. There is not one *nanotechnology*; it is a mass term and this observation will be examined at a later time. *Applied nanoscience* is what is happening today. *Nanotechnology* is an idea, maybe even a paradigm and a subject to be disputed later, so pardon my use of the clumsier term, *applied nanoscience,* but it seems more accurate in most illustrations that follow. When I use *nano-technology*, I mean to. To the questions then.

Who are the stakeholders? These would include those at the fore-front of the industry. Leading the pack are the scientists and engineers whose careers are linked to nanoscience. Next, we have the managers (including stockholders for public companies) through the company's or corporation's executives. On a different front, we have the insurance industry, whose job it is to protect the public interest by evaluating health and safety as liabilities. They are joined by the world of lawyer-ing, which gets to interpret claims made against producers who alleg-edly violate the public trust. Add to this mix some elected government officials and a small cadre of bureaucrats who promote nanoscience in terms of funding research, stimulating commercialization, and com-municating the public message. Add others as well who are in the front lines of regulating products of all sorts involving nanoscience. While there are a few public service interest groups who promote nanosci-ence, there are many more opposing globalization and/or promoting environmental considerations who argue against its advancement. There are some members of the media who write regularly about it and some members of academia (not unlike me) who find their tenures associated with nanoscience. Finally, but clearly not unimportantly, we get to the public who work in manufacturing and who consume prod-ucts made from the products of nanoscience.

Are the stakeholders equal? The answer is clearly "no." Those individ-uals and groups who directly benefit or suffer from the allegedly finan-cial windfall associated with the commercialization of nanoscience (if it occurs at all) understand that their interests at times are not the same as stakeholders who might approach the subject with more objectivity. That gets us to an assessment of public service interest groups, many of whom are dependent on public unease over technology policy for both political and financial capital. Who would donate money to protect the environment if it was not in trouble? As such, some groups are inter-ested in hyperbolizing the negative consequences of applied nanosci-ence and their remarks must be scrutinized. In addition, one need only procure a domain name from a subscription service, design a web page (and most of those services are bundled with the domain subscription), and behold—you have a public service interest group being run by a few guys with an axe to grind in the basement of their parents' homes. As such, some groups claim to represent the public interest without receiving any sanction from the public in any form whatsoever. This gets us to the public. Who is the public with an interest in science and technology policy making (a special definitional issue when limited to nanoscience)? In terms of national interest, the number of citizens interested in science policy is very low. We have learned few know much about nanotechnology per se. We also have observed efforts to reach out to the public have been woefully inadequate. To date, there remains no national clearinghouse and no source of information written for a public audience except for some inflammatory media reports, often written by journalists without any understanding of applied nanoscience. (My new book in preparation claims the media reports pornographically, much more concerned on stimulating interest and readership/viewership than true sustained attention and cognition). Few members of the public ever vote on any issue of science and technology policy, because they do not understand the subject, do not have time to learn more about the subject to have an informed opinion, and never get the chance. Some of my colleagues want to scientiate (to make scientifically literate) the public and improve their science education assuming that they will embrace the idea of nanoscience education, and upon learning more, will agree with the experts. Experience and recent data suggest otherwise. Hence, they defer responsibility for policy to elected officials and bureaucrats and we return to the beginning of the downward spiral.

Are we *really* interested in broad *stakeholder participation*? Absolutely not. Can you envision submitting science and technology policy to a national referendum? When studying the public sphere, we learned they were white male landowners. In more than one sense, they had a lot of capital. From Plato, to Habermas, to Dewey, they were elite and their justifications as protectors of the public interest run much like the case for the U.S. Senate in opposition to the more populist U.S. House of Representatives: to preserve the interests of minority views while advancing the general welfare. Put simply, the lay public (non-elite) outside this historical sphere can be fickle, and at other times, downright wrongheaded. While political philosophy and the tenets of democracy and republicanism suggest some level of participation is in order, that participation has been coded as representative and generally limited to rituals like elections, and maybe, polling experiments.

What does *stakeholder participation* mean and how can it be improved? In truth, it means those interested in promoting nanoscience want to be assured the public will not protest the use of public monies to fund research and some commercialization (this is especially important given the long time it takes to get a return on investment in this field). On another level, they want to be assured that a public protest and/or boycott movement does not surface to risk government and industrial investments. Finally, and most importantly, they want the public to consume products involving applied nanoscience in order to encourage more commercial investments in the field. It remains difficult to consider the *public sphere* in contemporary politics as much else than a consuming public. Indeed, we participate much more in policy by what we purchase than in whom we vote for every four years. If we really want to improve public participation, we need to find a way to involve them in a broad array of public outreach projects, from museum education, to deliberative polling exercises, to mass media documentaries. While some of these efforts are being undertaken, they remain exceptionally and woefully underfunded when compared to the millions spent to advance and promote applied nanoscience.

Finally, are some players using the rubric of *stakeholder participation* as a tactic to advance self-interests maybe even at odds with a broad public agenda? In 1964, Murray Edelman wrote a book on symbolic politics titled *The Symbolic Uses of Politics*. He claimed symbolic politics might be the bedrock of all political communications, but in the instance of nanotechnology, the degree and the intent of the symbology seems important. First and foremost, the public have already been self-selected from science literacy in most situations. For example, they decided not to major in chemistry and took the obligatory core lab courses. Subsequently, they are mostly lost when decoding claims and counter-claims in this area and have become nearly wholly dependent on others, especially the media. Another response by the public is to simply not attend to issues in this field unless they experience a personal epiphany, such as a significant other diagnosed with cancer or another disease. In nanotechnology, there is a similar public phenomena and most Americans know little or nothing about it and are not making much of an effort to learn more. While it is difficult to attribute motives to those who appeal for public involvement, there are some clues. The rhetoric from Washington, as it appears in report after report, tends to favor helping the public to understand as a step in promoting responsible nanotechnology. This rhetoric is also used by the few public service interest groups in applied nanoscience, such as the International Council on Nanotechnology, and in nanotechnology, such as the International Risk Governance Council. Most of the calls are couched in warrants associated with public support and the public sphere as consumers. Undoubtedly, a portion of the public attending to developments in applied nanoscience and nanotechnology must be asking whether being called a stakeholder and being generally asked to participate is anything more than a symbolic act obligated as a ritual in constitutional republicanism (our version of democracy). Those who participate in deliberate polling experiments, like consensus conferences and nano-juries, also ask whether their input will be taken seriously by policy makers. To date, it remains very difficult to say much more than "it should."

What precedes surfaced when writing *Nano-Hype: The Truth Behind the Nanotechnology Buzz* (Prometheus Books 2006), when posting on my blog (http://nanohype.blogspot.com/), while speaking at and attending many conferences here and abroad, and from reading nearly everything on nanoscience and nanotechnology policy that can be unearthed. While I do not want to appear sardonic or mordant, healthy skepticism in this arena may better protect the public interest. If we want public participation, let's have at it. If we believe the public are stakeholders or shareholders in this promotion, then let's empower them to participate meaningfully. If these hypothetical assumptions are not legitimate, then we should temper our rhetoric and get on with it.

 **David M. Berube** has been a PI or CoPI on four NSF societal and ethical grants totaling nearly $4 million. He has published dozens of articles in argumentation and in nano-science and technology risk and policy studies. He has written two books including *Nanohype: Beyond the Nanotechnology Buzz* (Prometheus Books 2006). He has degrees in psychology, biology, and communication, is a full professor in USC's graduate school, a member of the NanoCenter, and the government and industrial coordinator in the *nano*Science and Technology Studies program. In 2008, he will be moving to the Communication Department at North Carolina State University. He teaches graduate courses in risk communication, argumentation, and rhetoric in science and technology.

This work is supported by a grant from the National Science Foundation, NSF 01-157, Nanotechnology Interdisciplinary Research Team (NIRT)—Philosophical and Social Dimensions of Nanoscale Research, From Laboratory to Society: Developing an Informed Approach to Nanoscale Science and Technology and NSF 05-543, Nanotechnology Undergraduate Education (NUE)—Nanoscience and Technology Studies Cognate. All opinions expressed are the author's and do not necessarily reflect those of the National Science Foundation, the University of South Carolina, neither its NanoCenter nor its NanoSTS team, or North Carolina State University nor its Public Communication on Science and Technology (PCOST) Project.

11.3 The Context

The specific context in which the ethical agenda for nanotechnology is emerging is one of strongly mixed messages. There is general agreement, in theory, that the various societal concerns raised by technology need to be addressed in order to ensure public acceptance and support; however, the underlying intentions of various stakeholders may yet be strongly at odds (Berube 2006; Fisher 2005).[9,12] In addition, there are no real precedents for how to establish and implement a societal implications agenda. Increasing calls for a new field of "nanoethics" fail to address the fact that ethics generally is not a dominant force in science and technology debates. Simply sticking the term nano- on the front end is unlikely to alter this fact. What can be observed is that the same relatively narrow set of questions continue to be raised regarding ethics as part of the mandatory opening or closing statements of keynote speakers at societal implications conferences. Will nanoparticles be toxic? Will nanotechnology further impoverish developing countries? Should we allow human enhancement? However, there is no practical model for rigorous and effective ethical debate in the larger social sphere. Instead, what precedents we do have, such as the debates that rage in bioethics over the much narrower issues of abortion and physician-assisted suicide, are textbook examples of how not to conduct public dialogue.

It is probably safe to say we are faced with a genuine opportunity for social innovation. Several factors account for the dynamic possibility ahead of us. First, nanotechnology is not widely recognized or understood so there are few entrenched opinions to overcome (Macoubrie 2005; Cobb and Macoubrie 2004).[13,14] There is a proverbial blank slate from which we can experiment with new approaches to technology assessment and policy. Second, there is a small, but still unprecedented, level of funding available to directly address societal implications through the NNI. An infrastructure is emerging consisting of the various research centers funded by the NNI and charged with conducting societal implications research and public education as mentioned earlier. A third factor is the heightened pressure for and acceptance of interdisciplinarity in the scientific community that includes the realm of the social sciences and humanities. Novel collaborations are fertile ground for conceptual innovations and paradigmatic shifts in thinking. Finally, in the face of corporate scandals and government failures, there are increasing calls for public accountability from our institutions and a trend toward issue-oriented, grassroots activism such as the European experience with genetically modified organisms and the controversy in the United States over stem cell research.

11.4   The Stakes

The stakes are high and they involve everyone. Our ability to manipulate matter at the atomic and molecular level provides unprecedented power to shape our collective human future. Perhaps the most critical imperative we have is to find our way to an assessment process that balances our consideration of both the opportunities and the threats in both objective and subjective terms. In complex systems, a missed opportunity can quickly morph into an impending threat, and so the risks of failing to act when the opportunity is there can be as high as acting too quickly and without sufficient caution.

In the current environment, the stakes become even higher if we do not resist our natural inclination to oversimplify the systemic nature of the challenges. The stakes will be different for different stakeholders. They may be different in the short- and long-terms. The stakes may include very different outcomes in different parts of the system. Failure to view opportunities and threats with a strategic appreciation for the function of complex systems and the nature of uncertainty will leave us well short of meaningful options and choices.

At its core, what is at stake is a peaceful and prosperous future. Ethical priorities suggest at least a few of the larger goals to which any ethical agenda for nanotechnology should aspire:

- Protecting human dignity and fostering its highest potential
- Preventing avoidable harm and minimizing current and future human suffering
- Ensuring economic and environmental sustainability for current and future generations, and
- Enabling a more just distribution of resources and opportunity across the globe

---

11.5 The Role of Foresighting

As the length of time between discovery and commercialization shrinks, oversight as a methodology for checks and balances becomes less useful. There are so many serious SEIN (social and ethical implications of nanotechnology) concerns that the only alternative is foresight, and that demands speculation and a host of uncertainties that irritate scientists to no end.

—David Berube[9]

Foresighting is a contemporary version of scenario analysis, a product of post-World War II military planning that was introduced in industry in the early 1970s as a response to globalization and rapidly changing market forces (Schwartz 1991).[15] With a myriad of increasingly sophisticated tools, foresighting involves the generation of alternative scenarios to help identify current and future opportunities and their associated risks in the face of unstable and uncertain markets. For example, foresighting has become a common approach to generating global scenarios in ecosystem assessment and sustainability studies.[16] The basic goal is to "model, understand and shape the future to their advantage" (Wehrmeyer et al. 2002).[17]

COMMENTARY

A Word to the Nanowise: We Need Transparency and Accountability in Public Nanoscale Science and Technology

Nigel M. de S. Cameron

As this ground-breaking work goes forward, it is vital that NELSI projects be energetically developed with an immediate twofold intent. First, they need to assess the ethical, legal, and social entailments of new technological possibilities in order to anticipate problems and opportunities that may lead to regulatory, legislative, or other governance interventions. Second, although it is recognized that NELSI assessment is key to securing public confidence that a continuing, robust, and transparent critique of the technology is in hand, there is the wider consideration that NELSI projects serve as bridgeheads into the public's awareness of the societal significance of the technologies. This second prong of educating the public is especially critical if, as many observers believe is likely, the questions raised by developments in emerging technologies become a dominant (perhaps *the* dominant) theme of public and political debate in the 21st century.

Thus, NELSI should be seen as an exercise in strategic communication. Plainly, its goal must be to introduce the public to the facts and make the public aware of informed opinion in respect to these facts, and this must be achieved in a transparent and self-critical fashion—or else it will run the risk of being seen as mere propaganda by enthusiasts for the technology or those who wish to embrace it to develop particular applications, or both. That is to say, that, while some see NELSI as a means of embedding public support for the technology and others seek to use it to critique some, perhaps all, of its possibilities, a true, honest, unsensationalized, and grounded NELSI function is central to the success of democratically developing science policy in a century that will, on any accounting, witness vast shifts in our technological capacities. Far from being an add-on, NELSI is the very nub of democratic accountability.

It is vital that we do not take for granted the existence of any social consensus on the significance of nano as an agent for social transformation. If technology offers new social options, their implications should be made plain and submitted transparently to public scrutiny and democratic evaluation at the earliest possible stage. This emerges as a key responsibility of public officials charged with the funding and/or overseeing the technology, in collaboration with those engaged in NELSI research. Their special knowledge grants them unique insight into the likely direction of research and its social implications.

Therefore, public science officials must take great pains not to permit their enthusiasm for the technology in question, or their personal, political, philosophical, or wider social views to influence their discharge of the public trust and the manner in which they articulate the significance of their work.

Adapted from the final chapter of *Nanoscale: Issues and Perspective for the Nano Century*. Nigel M. de S. Cameron and M. Ellen Mitchell, eds. 2007. John Wiley.

Nigel M. de S. Cameron directs the Institute on Biotechnology and the Human Future and the Center on Nanotechnology and Society at the Illinois Institute of Technology, where he is Associate Dean and Research Professor of Bioethics. He is also President of the Center for Policy on Emerging Technologies (CPET), a new nonpartisan Washington-based think tank on science and technology policy.

He is co-editor of *Nanoscale: Issues and Perspectives for the Nano Century* (John Wiley, 2007), and has written widely on technology, policy, and ethics. He serves on the advisory boards of Nanotechnology Law and Business, the Converging Technologies Bar Association, and the World Healthcare Innovation and Technology Congress, and was co-chair of the 2005 International Congress of Nanotechnology. He has also represented the United States as bioethics advisor at the United Nations on biopolicy, and testified before the U.S. Congress and the European Parliament.

Foresighting methods are especially well suited to conditions of uncertainty. The concept of residual uncertainty was discussed in Chapter 1. Nanotechnology presents us with a classic case of this level of residual uncertainty. Nanotechnology has significant potential for technological, economic, and social disruptions, it is beginning to generate multiple new markets, and its full impact won't be predictable for several more years. Foresighting, by acknowledging a high level of uncertainty, forces consideration of a range of alternative futures and the anticipation of both optimistic and pessimistic outcomes.

At the same time, foresighting is also limited by the degree of uncertainty present in each scenario. The higher the degree of uncertainty, the less confident we can be that any of the scenarios is accurate. Nonetheless, Berkhout and Hertin (2002) contend that even though the future cannot be entirely predicted, future considerations can inform decision making in the present. They advocate interactive and participative methods to generate scenarios in order to ensure that the process captures both rational analysis and subjective judgment.[18] It is here we begin to see a role for ethics in the foresighting process. The ethical questions posed throughout this book can serve as one platform for both rational analysis and subjective judgment regarding assumptions about societal impacts in each scenario. Ethical issues can provide a critical measure against which uncertainty and its attendant risks can be weighed.

Alternatively, Davis (n.d.) is somewhat less optimistic about the ability of foresighting to adequately overcome uncertainty as a barrier to planning. He notes that "strategic planners are often faced with massive and ubiquitous uncertainty in many dimensions" and within systems that are "not only complex, but also adaptive."[19]

> To make things worse, when planners consider alternative strategies, they typically attempt to do so by predicting the consequences in future system behavior of their choices. However, they often discover—if they have the courage and integrity to address the issue—that they are unable to make meaningful predictions: future behavior of their complex adaptive systems is sensitive to a myriad of uncertainties.[19]

Clearly, the ability to generate accurate and useful scenarios is difficult; however, strategically informed anticipation remains the only really viable alternative to either excessive or inadequate caution. One of the great challenges of planning for the disruptive effects of nanotechnology will be the identification of critical response elements in the social infrastructure. Our ability to adapt to the societal effects of emerging technologies will rely, in part, on the effectiveness of key elements within the infrastructure and the extent to which they are sufficiently flexible and adaptive to respond to untoward outcomes. For example, regardless of the nature or source of harm, will regulatory and policy agencies have an effective process of rapid response that can address both suspected and completely unexpected harms? How quickly can economic safety nets be rolled out in the face of a severe economic disruption in one or more industries? Once again, ethics can prompt dialogue and provide guiding principles throughout the continuum from research and development to market release and regulatory oversight, at key decision points such as those discussed in Chapter 5. By posing the same ethical questions at such points along the way, we may be able to build responsiveness into the system ahead of untoward effects, even if the specific effects are not fully understood or anticipated.

We may not be able to control all of the details of how technology emerges and its subsequent outcomes; however, we can integrate ethical considerations at critical decision points in the process. We can control our intentions and we can set moral boundaries that favor the responsible development of nanotechnology. While it is clearly not practical for each nanotechnology research initiative to reinvent the wheel with respect to anticipating all possible societal outcomes of their research and development process, it may be quite practical to begin generating detailed scenarios using modified versions of the tools commonly used in business planning (Bennett-Woods 2007).[20] In the best case, such scenarios would be created through active partnerships between nano-researchers, engineers, social scientists, ethicists, business, government, and the public. They would then be available through commonly accessible databases that could also track actual outcomes via literature reviews and news stories. With broad participation at logical decision points, such as funding proposals, a centralized knowledge repository could be created on the basis of relatively simple, but continuous, forms of reflection, dialogue, and inquiry.

11.6   Ethics Applied to the Practical

There are three practical outcomes to a well-considered ethical analysis. The first is deep understanding of the values-driven issues at the heart of a problem to be solved or a decision to be made. This deep understanding then enables the second practical outcome, that of a well-reasoned solution or decision. The third practical outcome is the setting of a moral precedent or rationale that can be used for reference when considering new or related problems and decisions. When done effectively with the full engagement of key stakeholders, a fourth set of related outcomes is also possible.

In group settings, ethically grounded solutions and decisions can result in greater consensus and, therefore, stronger buy-in from those affected. Consensus building within the process of collaborative planning is a common means of strategizing among stakeholders when faced with controversial, uncertain, and complex conditions (Innes and Booher 1999).[21] The overt goal of consensus building is to bring together a representative mix of stakeholders and their interests to address issues of common concern using a form of mediated conflict or dispute resolution. Innes and Booher suggest that, in addition to the tangible result of an actionable solution, the most important outcomes of consensus building may be in the second and third order effects that result from "new relationships, new practices, and new ideas" generated in the process. They suggest that an effective process of consensus building can literally change the actors themselves, "moving a community to higher levels of social and environmental performance." Specifically,

> A process that is inclusive, well informed, and comes close to achieving consensus is more likely to produce an implementable proposal than one lacking in these qualities. If it follows principles of civil discourse, it is more likely to build trust, foster new relationships, and create shared learning. If it encourages participants to challenge assumptions, it is likely to produce new ideas. Stakeholders are more likely to feel comfortable with a process they can organize themselves and more likely to be committed to its results.[21]

Likewise, constructive ethical discourse embodies all of these qualities. By using a variety of ethical lenses, the process is more likely to include a representative range of viewpoints and concerns. It will also lead to a more widely informed analysis as questions raised require a broader and more detailed description of the relevant context. Trust and relationships are built on a foundation of shared values, mutual goals, and jointly perceived insights. The challenging of moral assumptions with compelling counterarguments forces dialogue to a more creative depth as participants work to resolve problems of consistency and coherence in their views. The goodwill generated through common understanding and respect results in a commitment to meeting multiple needs and avoiding needless harms.

In fact, although promoting a general communications strategy, Innes and Booher also raise subtle ethical questions in their assessment of the process. They posed three key questions. Would the outcomes be just and fair? Would they be sustainable? Would they be, in some sense, for the common good? In response, they point out that wide representation of stakeholders is inherently likely to produce a more just outcome as well as serving a shared conception of the common good. Furthermore, a process that fully explores the options and consequences of actions is likely to generate more sustainable solutions as well as building the capacity to adapt creatively to change over time. Interestingly, they captured the essence of the principles of community, the common good, and social justice proposed throughout this book.

11.7   An Action Plan

The conscious integration of an ethical focus in all corners of the technology assessment arena is no small task. There are any number of barriers including:

- Lack of interest
- Lack of will
- Lack of understanding
- Lack of process
- Lack of a common language or understanding

It is notable that the voices missing from the commentaries in this book are the voices of government and business, with one notable exception. It is not because they were not invited. They were. Invitations were either ignored or declined on the grounds the invitee was not qualified to comment, too busy to comment, or simply critical of efforts to discuss nanotechnology in ethical terms.

Despite the barriers, there are significant opportunities as well. While nanotechnology may present a unique opportunity for building a more

COMMENTARY

Bruce V. Lewenstein

To say that there are social and ethical issues in nanotechnology is to say that science and technology exist only in a social context. Better public understanding of the interaction of science, technology, and society will, many of us believe, lead to more informed decisions about how to invest in science and technology, when and how to regulate—or not regulate—technological development, how to address inevitable ethical challenges, and so on.

But what does "better public understanding" mean? Significant research in the area of public understanding over the last twenty years has shown that simple definitions of "science literacy" or measures of public knowledge are far too limited in their conception of public understanding (Gregory and Miller 1998; Irwin and Wynne 1996; Miller 2001).[1,3,6] Instead, the idea of "public engagement" in science and technology has become the preferred way of thinking about the interaction of science, scientific knowledge, scientific institutions, and their many audiences (House of Lords 2000; Leshner 2003).[2,4]

The challenge is to recognize that "public engagement" is not just a different set of words, but a fundamentally different idea. When scientists, science journalists, and others started using the words "public understanding of science" in the 1940s, what they actually meant was "public appreciation of the benefits that science provides to society" (Lewenstein 1992).[5] While I don't disagree with that idea, I think it's important to recognize that better understanding may not yield more appreciation or support. There is substantial evidence, for example, that those among the general public who pay more attention to science than the average college graduate have a *less* positive attitude towards supporting science (National Science Board 1996, p. 7.18).[7] Public engagement, on the other hand, is about two-way dialogue, about listening to each other, about truly engaging in each other's concerns, about shifting one's claims and position—not simply about persuading someone to agree with your original position (Wilsdon and Willis 2004).[9]

Taking this approach to public engagement, however, means recognizing that in a democratic system, the power to make decisions rests with the citizenry, not with the experts. To truly participate in public engagement or public participation, scientific experts must recognize that their judgments about, for example, the acceptable risk-benefit ratio of developing nanomaterials, may not be the same judgments as those of broad public *informed* groups that consider the same issues.

How is this an ethical issue? Because, like issues of privacy, intellectual property, environmental risk, workforce development, or human enhancement (other topics often labeled as "social and ethical issues" in nanotechnology), public engagement is ultimately a question of power in social relationships. In each case, not only are legitimate questions possible about how nanotechnology research and application should develop, but even more fundamental questions exist about how to make decisions and who should control those decisions. These fundamental questions are asking about the source of power in societies with unequal social distributions of power.

Though scientists often complain about what they perceive as a lack of social power, they are in fact one of the most respected social groups in society and their judgments are highly regarded (National Science Board 2006).[8] In the United States, in particular, "expertise" is a valuable social resource, and in times of political conflict (such as in debates about nanotechnology) competing groups fight to claim the mantle of "science."

Thus the ethical challenge in exercises of public participation in nanotechnology is to manage not only what people know and how values are expressed, but also who will make and enforce the decisions about nanotechnology—and all other emerging technologies—that society must make.

References

1. Gregory, J. and S. Miller. 1998. *Science in Public: Communication, Culture, and Credibility*. New York: Plenum.
2. House of Lords. 2000. *Science and Society*. London: UK House of Lords.
3. Irwin, A. and B. Wynne, eds. 1996. *Misunderstanding Science? The Public Reconstruction of Science and Technology*. Cambridge: Cambridge University Press.
4. Leshner, A. I. 2003. "Public Engagement with Science." *Science* 299, 977.
5. Lewenstein, B. V. 1992. "The Meaning of 'Public Understanding of Science' in the United States after World War II." *Public Understanding of Science* 1(1):45-68.
6. Miller, S. 2001. "Public Understanding of Science at the Crossroads." *Public Understanding of Science* 10(1):115-120.
7. National Science Board. 1996. *Science and Technology: Public Attitudes and Public Understanding. In Science and Engineering Indicators—1996* (Chapter 7). Washington, D.C.: U.S. Government Printing Office.
8. National Science Board. 2006. *Science and Technology: Public Attitudes and Public Understanding. In Science and Engineering Indicators—2006* (Chapter 7). Washington, D.C.: U.S. Government Printing Office.
9. Wilsdon, J. and R. Willis. 2004. *See-Through Science: Why Public Engagement Needs to Move Upstream*. London: DEMOS.

 **Bruce Lewenstein** is Professor of Science Communication in the Departments of Communication and of Science and Technology Studies at Cornell University, Ithaca, New York. He works primarily on the history of public communication of science, with excursions into other areas of science communication. He has also been very active in international activities that contribute to education and research on public communication of science and technology, especially in the developing world.

He is a co-author of *The Establishment of American Science: 150 Years of the AAAS* (Rutgers Univ. Press, 1999, with Sally Gregory Kohlstedt and Michael M. Sokal), editor of *When Science Meets the Public* (Washington, D.C.: AAAS 1992), and co-editor of *Creating Connections: Museums and the Public Understanding of Research* (Altamira Press, 2004, with Dave Chittenden and Graham Farmelo). From 1998 to 2003, he was editor of the journal *Public Understanding of Science*. He is a Fellow of the American Association for the Advancement of Science.

strategic approach to science and technology assessment, there are any number of other societal challenges that would benefit from more robust formats for civil dialogue that incorporate a reasoned approach to questions of value and meaning. Although it can certainly be used poorly, ethics and ethical analysis is not inherently dogmatic, prescriptive, or inflexible. If you are going to label a problem ethical, then you need to approach it from the ethical dimension. At this point, there seems to be more willingness to call something an ethical issue than to actually analyze it as such. So, how do we move from simply pointing out ethical problems to actually talking about them?

11.7.1 Ethics 101

The first item in the action plan may be to rehabilitate the reputation of ethics itself. As with any academic discipline, the scholarly pursuit of ethics can indeed be theoretically dense and inaccessible to lay persons, at least in any practical way. If philosophers and ethicists want to be included in the conversation, then they need to emerge from behind their scholarly journals and speak a language everyone can understand and engage in. Anyone with a basic understanding of the underlying assumptions can apply basic ethical concepts to a problem. The goal of those of us who teach or otherwise do ethics for a living should be to use our skills with the language and process of ethics to facilitate the dialogue in the broader social sphere.

11.7.2   Create Forums

The second action item will be to collaboratively make use of existing forums and create innovative new forums for ethical dialogue. Incorporating a dedicated ethics course into science and engineering courses is an obvious first step. This includes programs in the social sciences, which don't generally require a course in philosophical or applied ethics beyond perhaps the ethics of human subjects research. Such a course must go beyond simple prescriptives such as the introduction of professional codes of ethics (Haws 2004).[22] Rather, it should include a robust component of theory while focusing on the application of theory in practical decision making.

Similar forums for basic education should be envisioned within government and industry. For example, most hospitals in the United States have an established ethics committee consisting of interested employees who receive ongoing education in bioethics, and then serve in an advisory and consultative capacity within the organization to help resolve difficult policy issues and patient care dilemmas. They do not have decision making authority per se. Similarly, many hospitals and all universities have an Internal Review Board (IRB) that oversees human subjects research. In both cases, individuals with an acquired expertise become organizational assets for ethical analysis and decision making. Similar entities could be used within government and industry to assist in the review of funding proposals, decisions to transfer knowledge, formulation of policies, public relations strategies, marketing, and, most importantly, support of a culture of ethical awareness within their respective organizations.

There is much work being done on the creation of forums for public engagement. For our purposes here, the particular format and approach is not as important as how it deals with the ethical dimension. What is important is to formulate questions and prepare facilitators in a manner that encourages genuine engagement with ethics. If questions or scenarios are presented in narrowly utilitarian terms, then the result is likely to be a narrowly utilitarian response.

A further innovation would be to incorporate ethical dialogue and analysis into the very fabric of the research process. Following calls for upstream engagement of societal issues, this highly applied forum is the logical goal to which we should all aspire. Every effort that separates ethics from the day-to-day decision making creates an artificial boundary that can be easily dismissed as just another bureaucratic hoop. Efforts to integrate societal implications assessments into a more real-time assessment process are in their infancy (Guston and Sarewitz 2002).[23] Again, the tactics and methods of accomplishing the broad goals of a more informed and proactive technology assessment may vary widely and still be effective from the standpoint of ethical implications so long as they consciously incorporate the ethical dimension with an element of rigor and depth.

11.7.3   Establish a Common Language

One of the primary challenges faced in the consideration of nanotechnology generally is the lack of common definitions and terms. These largely technical terms will eventually work themselves out as the marketplace and regulatory bodies push for standardization. Ethical dialogue suffers from a similar problem when it comes to the definition of concepts. My definition of justice is probably different than your definition, at least in how we might apply to a specific case. Nonetheless, there is great value in debating how we define and apply ethical concepts and we can eventually achieve a certain level of consensus that allows us to more effectively frame questions, identify common concerns, debate boundaries and assess solutions.

As such, on a societal level, ethics is very much an iterative dialogue in which stakeholders define, redefine, and ultimately refine key concepts such as basic human rights, social justice, the greater good, harm, and duty. As the dialogue unfolds, it will be important for scholars to track its trajectory and synthesize whatever consensus does begin to emerge on the most pressing questions, the concepts being debated, the actions suggested, and any identified barriers to resolution and action.

Inevitably, some issues will end up with calls for legislative action and regulation; however, only rarely is this a solution to deep moral dilemmas. Ethics and the law have an uneasy partnership in which there are many examples of laws that are fundamentally unethical and unethical actions that are technically legal. Regulation tends to paint solutions with a broad brush that end up causing harm or unfairness to at least a portion of those it involves. Regulation rarely reflects a complete understanding of the front line at which it must be implemented. Regulation is not particularly responsive to a rapidly changing or complex environment. Finally, legislative responses to fundamental social issues have proven to be political minefields that seesaw back and forth depending on who happens to be in power at the moment. For these, and other reasons, legislative solutions should generally be a last resort, used only when it is clear that science, government, and/or industry are unresponsive to the valid ethical concerns of their constituencies. In these cases, once again, well-formulated ethical questions can be useful in informing the policy and legislative process in a meaningful way.

11.8   Citizenship in the Nano-Age Revisited

The profound social and political challenge of our time is how to quell the increasing distrust, alienation, and disengagement felt by citizens who feel powerless in the face of unresponsive social institutions. Ethics, and the concept of the moral agency of each individual person, naturally holds us accountable to ourselves and to each other. Whether we are naturally inclined to consider a utilitarian greater good or a communitarian version of the common good, as citizens we are obligated to make choices that allow us to coexist in community. We have a duty not to cause or contribute to needless or avoidable harm, be it through direct action or simply failing to act in the face of potential harm. We have a duty to act in the best interests of ourselves, our family, and our neighbors. Furthermore, we have an obligation to do so with the awareness of a shrinking world in which we can no longer afford not to consider people halfway across the globe as neighbors in every sense of the word. Such global awareness calls us to a duty of justice in which we strive to define what is equitable and fair in a way that broadly enables human potential for more than just the fortunate few. We have a duty to be accurately informed, in order to seek a common truth beyond the political, religious, and cultural fault lines that threaten the future well-being of the entire human community.

Ultimately, the inherent dignity of the human species lies in our ability to reflect, develop insight, and act with wisdom. The failure to live with integrity and intention represents an undermining of human potential. Science and technology are not the inevitable ends of human existence but byproducts of the larger human potential. They will only be able to serve that human potential as well as we allow them. Responsible and effective citizenship is, and always has been, a function of the choices we make. Even the most superficially practical choice reflects an underlying structure of values and value-based assumptions.

11.9   Questions for Thought

1. How does your organization address ethical issues? Is there a mechanism for reporting ethical concerns or a forum for discussing the ethical implications of what you do?

2. Select an example of a nanotechnology application that may prove controversial in some way. Imagine how the issue might play out in the funder, thinker, and communicator arenas. How will each arena influence the larger societal response? Which arena will be the most effective and which will be the least effective in achieving a justified solution to the issue.

12

Reflections on Technology and the Moral Imagination

12.1 Why Bother?

Science is organized knowledge. Wisdom is organized life.

—Immanuel Kant

The writing of this book has been an interesting, albeit challenging, journey. Had I fully appreciated the scope and breadth of the content, or the distance from my normal comfort zone in health care and bioethics, I would have had to think long and hard about agreeing to write it. I entered this project without a strong position on any of the topics I intended to cover, and I worked hard to maintain the same even-handed treatment throughout the book. Now, poised at its conclusion, I feel compelled to violate the mantle of objectivity to offer a personal reflection. The intention of this last chapter is to give voice one last time to the essence of ethics as a practical and trusted companion as we move into our technological future.

When confronted with multiple, opposing viewpoints and compelling, but conflicting arguments on difficult topics, my students often initially throw up their hands and adopt an attitude of, "Why bother?" The issues are too complex, the questions too hard, the control too far out of their hands, and the consequences appear to be inevitable no matter what they think or do. I then must spend the rest of the course undoing the "damage" I have done when challenging them to think deeply and from many points of view. I must rebuild their confidence and encourage an attitude of personal engagement and professional leadership in their own lives. Leadership is not a position or a title, but rather a way of being in the world that is grounded in a deep self-awareness, and a sense of personal and professional purpose. Alternatively, the attitude, "Why bother?" is the antithesis of both leadership and ethics. It is profoundly contrary to the demands of professionalism and the very foundations of a free, robust, and self-sustaining democratic society. Each individual's failure to engage in the moral reflection needed for professional accountability and informed citizenship is a nail in the coffin of the peaceful, prosperous, and sustainable future most of us seek for ourselves, our children, and our grandchildren. "Why bother?" kills innovation and creativity; dulls personal responsibility and concern for others; suffocates optimism; and, breeds apathy and inaction. It strips both meaning and life from a community.

I was taken with an essay by Bruce Grierson (2007)[1] that appeared in *Time* magazine earlier this spring. Grierson is the author of *U-Turn: What If You Woke Up One Morning and Realized You Were Living the Wrong Life?*, a clever book about people who suddenly adopted beliefs nearly opposite those they had previously held, and altered their life trajectories accordingly. When describing these epiphanistic turns, he says, "I realized this was what almost all the U-turns had in common: people had swung around to face east. They had stopped thinking in a line and started thinking in a circle. Morality was looking less like a set of rules and more like a story, one in which they were part of an ensemble cast, no longer the star."

Earlier in the book, I stated that ethics and morality are more than just collections of abstract theories and principles. Grierson captures my point explicitly. Ethics is the story of how we, as individuals and communities, ultimately choose to unfold. An ethic of the practical, an ethic of the ordinary, accomplishes exactly what Grierson describes. It hovers in the silent corners of our personal reflections and embeds itself within even our most mundane conversations and decisions as our individual and collective stories come into being. Ultimately, it composes the frames in which the snapshots of our common existence are captured, displayed, and remembered. Our personal and communal stories are always told in terms of the good we did or did not do; the harms we did or did not have the foresight to avoid; the dignity and humility with which we did or did not face adversity; the courage we did or did not have to stand up for our beliefs; the compassion we did or did not show for others; and, the wisdom with which we did or did not use our power.

Within those stories, modern technology is a dominant force that molds our experience of the world and each other. The forms it takes and the purposes it assumes are ultimately a reflection of the values we have assigned to it. Nanotechnology and its convergence with other emerging technologies may radically alter our relationship with each other and the world around us. The ability to manipulate matter at the atomic scale, to model life itself in our machines, and to incorporate our technology into the innermost reaches of our own bodies represents a power that should not be taken lightly. The wisdom we apply to how we ultimately decide to use or not use this power matters deeply. It will shape our world and is definitely worth the bother of an ongoing conversation.

12.2 The Value of the Skeptical Optimist

A pessimist sees the difficulty in every opportunity; an optimist sees the opportunity in every difficulty.

—Winston Churchill

The story of our unfolding has always been populated with a diverse range of characters, including the profoundly skeptical and the unabashedly optimistic. The subject of technology is one in which this diversity of temperament is particularly apparent. Skeptics and pessimists often fear technology, question its value, and doubt our ability to control its negative or untoward consequences. They can be overly quick to recognize the inherent risks and harms, and overly cautious in weighing such harms against the benefits. They often place little trust in institutions, and even in people in general. They can, at times, resist change more as a matter of principle and habit than reasoned analysis.

Optimists, on the other hand, can be equally persistent in their insistence that "it's all good." They often embrace technology with an eagerness that borders on the reckless. They never question its value and see little need for control, believing that problems will simply work themselves out as we go along. They can be overly quick to dismiss even the most reasonable risks and harms, always assuming the benefits will far outweigh any harm that might arise. They often embrace change for its own sake, trusting in the ideal of progress as a good in and of itself.

Skeptics and their pessimistic leanings can suck the energy out of just about any endeavor. Relentless optimists are equally wearying in the other direction. Both lack true moral imagination, and the ability to step into the values, assumptions, and perceptions of the other. They personalize, polarize, and politicize the dialogue far more than they facilitate it. At their worst, they are incapable of seeing any merit, giving any credence, or even bothering to actually listen to the positions of those whose views diverge from their own. Finally, both skeptics and optimists can be unashamedly self-interested in their intentions.

Yet, if we are to exercise wisdom in our pursuit of scientific mastery and ever more powerful technologies, we would benefit greatly from developing an appropriate moral imagination to accompany us on that journey. Perhaps the simplest example of moral imagination is found in the many versions of the Golden Rule. Do unto others as you would have them do unto you. Found in faith and wisdom traditions across the globe and across human history, this deceptively simple concept embodies a singular approach to the moral dimension of living in community. It involves contemplating the impact of my actions by literally stepping into the viewpoint of every other person through my own eyes and insights. First, I am called to imagine how I would feel and respond if someone treated me in the way I am about to treat this person. Then, I am forced to take the position of the other, and imagine an action that treats this person with the same dignity, respect, and care that I would wish for myself.

The well-developed moral imagination naturally asks hard questions as a matter of habit. What harm will this cause and to whom? Is it fair? Is it caring? Does it unite or does it divide? Such imagination then goes about answering those and other questions, perhaps not with objective precision, but with a realistic understanding of human fallibility and a focus on our better human instincts and higher aspirations. Herein lies the value of the skeptical optimist. These are people who believe deeply in the essential drive and resilience of the human heart and mind to reach beyond itself in its search for goodness, truth, and wisdom. They recognize that we are all more alike than different and, as a species, we are inherently capable of acting in the best interests of ourselves and others. However, these are also persons who recognize our human limitations. They admit that our tremendous capacity for discovery and control often exceeds our ability or will to foresee and understand the broader implications of those same discoveries, and the power that underlies control. They deeply comprehend the difficulties in balancing the interests of self and community, as well as anticipating short-term and long-term effects. The optimist in them can genuinely appreciate the positive vision, while his or her inner skeptic can simultaneously be patently forthright about its inevitable downside and our history of often not looking before we leap.

12.3   Pace, Complexity, and Uncertainty Revisited

> The march of science and technology does not imply growing intellectual complexity in the lives of most people. It often means the opposite.
>
> —Thomas Sowell

Economist Thomas Sowell expresses a worrisome disconnect between what we create and our ability to understand it. An overriding theme of this book has been the mandate to consider the escalating pace of discovery and change in an increasingly complex and uncertain environment. Each of these conditions constitutes a fuzzy boundary across which the many questions raised in this book must be negotiated. Each has a dark side, and each a light side. Escalating pace dramatically reduces time for reflection, testing the waters, achieving full understanding, and building consensus. The dangers inherent in moving too quickly ahead of the dialogue cannot be overstated given the rapid diffusion of new technologies into widespread use, and the almost virtual impossibility of putting them back into the proverbial box. On the other hand, serious problems can magnify quickly and we want promising technological advancements that provide solutions to be able to move quickly through the research and development process without undue restriction.

Complexity and uncertainty work hand in hand to complicate any reasoned analysis of the potential impacts of nanoscience and nanotechnology. The more complex and uncertain the context of a decision, the more time we need to sort it out—yet the less time we have in the face of increasing pace. Our comfort zone lies largely in the linear models of thinking and organizing that have served us reasonably well in the past. However, these simple cause and effect models tend to break down in the face of complex systems and high levels of ambiguity.

In his classic text on the learning organization, Senge (1990)[2] urges us to learn how to think in terms of both detail and dynamic complexity. Detail complexity involves many different variables at work in the situation, requiring us to identify and measure multiple variables at once. Senge suggests that high levels of detail complexity render most explanations incomplete at some level; however, he also points out that the human brain is generally pretty good at handling detail complexity. I would add that information technology has greatly enhanced our capacity for detail complexity as well. In fact, a computer program is itself a good example of detail complexity. A conventional program with a million lines of code has a tremendous amount of detail complexity; however, finding the bug in one or two lines of code is a basic cause and effect process. It may take time, but the process is essentially a linear one.

Dynamic complexity, on the other hand, is an effect that involves situations in which "cause and effect" are not always apparent. Cause and effect relationships may not be obvious and may be separated in time and space. Even more challenging, the relationships may produce utterly unexpected outcomes including one outcome in the short-term and quite a different outcome in the longer term, or one outcome in one part of the system and another outcome in another part. Understanding situations in terms of dynamic complexity requires systemic thinking and non-linear pattern recognition that includes identifying feedback loops and recursive or iterative systems effects. For example, the fact that materials at the nano-scale have markedly different properties than the same materials on the bulk level gives them a form of dynamic complexity that makes it very hard to anticipate their impact on the environment based on prior knowledge or past experience. Likewise, human cognitive enhancements represent a tremendous level of dynamic complexity, making it very hard to anticipate the physical, emotional, social, and cultural impacts such technologies might engender on both the individual and society as a whole. Citing examples such as climate change, nuclear proliferation, poverty in developing nations, and economic cycles, Senge points out, "The human being is exquisitely adapted to recognize and respond to threats to survival that come in the form of sudden, dramatic events. Clap your hands and people jump, calling forth some genetically encoded memory of saber-toothed tigers springing from the bush. Yet today the primary threats to our collective survival are slow, gradual developments arising from the processes that are complex both in detail and in dynamics."

Notice that the term slow is relative here when considering the accelerating pace of change. Nanotechnology is not likely to suddenly arrive, rather it is and will continue to emerge in waves of discovery and application that shift our capabilities over time. It may, however, be a very short period of time in contrast to prior human technological development.

When discussing organizational dynamics, Senge suggests the real leverage in most management situations lies in understanding dynamic complexity rather than detail complexity. It follows that the unfolding of nanotechnology within multiple technical and scientific disciplines, and simultaneously across many economic sectors, will present us with a great deal of detail complexity through which to wade. More importantly, it will also present a whole new level of dynamic complexity. Our ability to anticipate and then connect the dots between various societal impacts will largely determine how effectively we respond to the many ethical questions raised in prior chapters of this book.

Finally, the uncertainty surrounding nanotechnology and its convergence with biotechnology and other disciplines will remain; persisting despite our best efforts to control pace and unravel complexity. Two responses will be needed. The first response is to minimize uncertainty as best we can via the development and use of increasingly sophisticated methods of forecasting and risk assessment. These methods must address both the scientific and technological impacts as well as the societal impacts. The second strategy is to call upon the social sciences and humanities to investigate the technological impacts on society in ways that can guide the direction of nanotechnology policy. Our goal should be to engage a well reasoned and appropriately measured societal response to nanotechnology as far in advance as possible. Framing the ethical questions is a direct means of identifying the underlying uncertainties for which we must plan if we are to use nanotechnology and its enabling power wisely and well.

12.4   Final Thoughts

> Even bigger machines, entailing even bigger concentrations of economic power and exerting ever greater violence against the environment, do not represent progress: they are a denial of wisdom. Wisdom demands a new orientation of science and technology towards the organic, the gentle, the nonviolent, the elegant and beautiful.
>
> **—E.F. Schumacher**[3]

Economist E. F. Schumacher's[3] thoughts contain a double meaning here at the end of this book. On one level, the comment can be read as a warning against the destructive power of science and technology imposing itself in a manner that allows it to stray too far from that which best serves the limits of our natural environment, as well as our human nature, individual spirit, and community. On another level, it implies the constructive power of a science and technology that mimics the elegance of nature's own bottom-up process of creation and generation. The promise of nanotechnology is that it will give us unprecedented power to manipulate our environment. The peril of nano-technology is that we will lack the wisdom to use that power well.

Spend a few hours in the societal implications literature related to nan-otechnology and you will find no shortage of questions raised, but few answers. Likewise, many questions have been raised in this book, along with suggestions for various ways on which we might think about possible answers. In an exploration of professional ethics in science, Chadwick (2005)[4] poses three basic philosophical questions attributed to the German philoso-pher Immanuel Kant. She observes that we have spent a good deal of time and energy trying to answer the first two with respect to science and with increasing focus on emerging technologies: what can I know and what ought I to do? She closes the paper with a suggestion that more attention be devoted to the third of Kant's questions: What may I hope for? I heartily agree. The question of what we can know is bounded by our tools and by our abilities to observe and reflect deeply on what we do or do not find. The question of what ought we to do is bounded, in part, by our values and system of ethics. However, this second question also presupposes a clear vision of our larger purpose, that for which we might hope.

Too often our "hopes" are overly narrow, self-interested, and contradic-tory. We hope for both longer, healthier lives and intergenerational eco-nomic justice without recognizing there may be an inherent and potentially unresolvable conflict between the two. We hope for economic success and material comfort as well as a clean, sustainable environment; yet, we tend to disregard the impact our material culture has on the uneven distribution of the world's scarce resources, and the environmental degradation that is part and parcel of our preferred lifestyle. We hope for safety and security without fully appreciating the implications for our privacy and other civil liberties, not to mention the potential for advanced weapons and defensive systems to fall into the wrong hands.

All successful enterprises are built on a preferred vision of the future—that for which we can hope. The Chinese philosopher Lao Tzu (Mitchell 1988)[5] advises us to "prevent trouble before it arises. Put things in order before they exist." Asking the hard questions first allows us to anticipate the strate-gies and safeguards that ultimately lead to the greatest benefit at the least cost or harm. Cultivating a vision of technology in which we are consciously prioritizing its value and our needs in a coordinated, strategic, and sustain-able manner may actually slow the process down a bit, but is likely to real-ize lasting benefits sooner and with less need for damage control on the fly. How can we best focus our investment of scarce resources? How can we maximize the benefits and minimize the harms? Is there a new and more effective paradigm of science and technology, one in which competition and collaboration work hand-in-hand to spur innovations that mitigate the existing economic, political, and social imbalances that lead to increased injustice and instability in an ever smaller world? And perhaps most importantly, how shall we define human potential as we move into the coming nanocentury?

12.5 Questions for Thought

1. Brainstorm a list of social goods that can be achieved, in part, with the aid of nanoscience and nanotechnology. This list might include human health, education, environmental protection, economic wealth, safety and security, etc. How would you prioritize them if asked to allocate science and technology funding for the next 20 years? Are there things you would not fund at this point? Frame your justification in ethical terms?

2. Is the current process of technological innovation sustainable? Is it even possible or desirable to direct it more strategically?

3. How would you define human potential in the face of nanotechnology and its eventual convergence with other powerful technologies in the next 20-30 years?

Glossary of Terms in Ethics

applied ethics: ethical rationales employed in the analysis of specific, practical issues and based on normative ethics.

autonomy: the ability to act so that your actions are the result of your own deliberation and choices. May be applied to individuals or communities.

beneficence: An obligation to act in ways that promote good, remove harm, and prevent harm.

compensatory justice: compensation for wrongs or harms that have been done.

deontology: (from the Greek *deon*, meaning "duty") refers to an ethical theory or perspective based on duty or obligation. A deontological, or duty-based, theory is one in which specific moral duties or obligations are seen as self-evident, having intrinsic value in and of themselves and needing no further justification. Moral actions are evaluated on the basis of inherent rightness or wrongness rather than goodness or a primary consideration of consequences. Kantianism, divine command theory, and some rights-based theories are generally categorized as deontological theories.

descriptive ethics: describes the ethical beliefs, norms, and behaviors of an individual or group as they actually exist, as opposed to how they ought to exist.

distributive justice: an equitable balance of benefits and burdens with particular attention to situations involving the allocation of resources.

ethical absolutism: the view that moral rightness and wrongness exist independent of human beings and are unrelated to human emotions and thought. There is an absolute source of truth that transcends human rationality and choice.

ethical relativism: holds that judgments about the rightness or wrongness of an act can legitimately vary between persons or cultures based on individual feelings (subjectivism) and specific social and cultural circumstances (cultural relativism). This view assumes that morality depends on a dual consideration of human nature and the human condition with specific social and cultural circumstances playing a role in determining moral beliefs and practices.

ethics: the branch of philosophy that studies morality through the critical examination of right and wrong in human action. Also termed moral philosophy or the science of morals.

fidelity: principle that broadly requires us to act in ways that are loyal, including keeping our promises, doing what is expected of us, performing our duties, and being trustworthy. Role fidelity entails the specific loyalties associated with a particular professional designation.

justice: principle that requires us to act in ways that treat people equitably and fairly. Actions that discriminate against individuals or a class of people arbitrarily or without a justifiable basis would violate this basic principle.

metaethics: concerned with the very nature of right and wrong, where and how ethical judgments originate, and what they mean in relation to human nature and conduct.

moral objectivism: holds that at least some moral principles and rules are objectively knowable on the basis of observation and human reasoning.

morality: the customs, principles of conduct, and moral codes of an individual, group, or society.

natural rights: moral claims that are generally held to be a gift of nature or God that cannot be taken away. Modern notions of natural rights have come to be referred to as universal human rights and form the basis for establishing and/or evaluating ethical standards within the social order.

nonmaleficence: principle that requires us to act in ways that do not inflict evil or cause harm to others. In particular, we should not cause avoidable or intentional harm. This includes avoiding even the risk of harm.

normative ethics: seeks to define specific standards or principles to guide ethical conduct in answer to questions such as what is valuable and how are actions morally assessed and justified.

paternalism: is the principle that allows a physician to act contrary to a patient's wishes if there is evidence that the patient is not acting in his or her own best interests and on the basis of a higher level of expertise.

principle: refers to a basic truth, law or assumption. In ethics, a generalization that can be used in moral reasoning or a specific rule of good conduct.

principlism: ethical theory and method in which each principle represents a serious, though not absolute, moral duty that must be weighed against other duties in resolving an ethical conflict or dilemma.

procedural justice: principle that requires processes that are impartial and fair.

respect for persons: principle that maintains human beings have intrinsic and unconditional moral worth and should always be treated as if there is nothing of greater value than they are.

right: a justified claim that can be made against another individual or group. In the case of a legal right, the claim must be justified by legal principles and rules. Likewise, a moral right must find grounding in moral principles and rules. One form of rights does not necessarily lead to another.

teleology: (from the Greek *telos*, meaning goal or end) describes an ethical perspective that contends the rightness or wrongness of actions is based solely on the goodness or badness of their consequences. In a strict teleological interpretation, actions are morally neutral when considered apart from their consequences. Ethical egoism and utilitarianism are examples of teleological theories.

universalism: suggests that basic right and wrong is the same for everyone, while also allowing for some variation in individual circumstances and context.

utilitarianism: a moral theory that defines a moral act solely in terms of the outcome or consequences of that act. This teleological perspective is based on a single guiding principle. The principle of utility, also referred to as the Greatest Happiness Principle, states that actions are right if they produce the greatest balance of happiness over unhappiness.

veracity: is the principle of truth telling.

virtue ethics: an approach to ethics in which the focus is on the role of character as the source of moral action. Human character is shaped over time by a combination of natural inclinations and the influence of such factors as family, culture, education, and self-reflection.

Source: Bennett-Woods, D. 2005. *Ethics at a glance.* Online resource. Available at http://rhchp.regis.edu/HCE/EthicsAtAGlance.

References

CHAPTER 1

1. Tolkien, J.R.R. 1965. *The Lord of the Rings, Part One: The Fellowship of the Ring*. New York: Ballentine Books.
2. Schick, Theodore. 2003. "The cracks of doom: The threat of emerging technologies and Tolkein's rings of power." (In G. Bassham and E. Bronson, eds.) *The Lord of the Rings and Philosophy: One Book to Rule Them All*. Chicago: Open Court.
3. Schummer, J. 2005. "Societal and ethical implications of nanotechnology: meanings, interest groups, and social dynamics." *Techne: Research in Philosophy and Technology* 8(2). Retrieved online June 4, 2007 at http://scholar.lib.vt.edu/ejournals/SPT/v8n2/schummer.html
4. Hatch, R. n.d. The Scientific Revolution: Definition, Concept, History. Retrieved May 8, 2007 from http://web.clas.ufl.edu/users/rhatch/pages/03-Sci-Rev/SCI-REV-Teaching/03sr-definition-concept.htm
5. Roco, M.C. & W.C. Bainbridge, eds. 2001. Societal Implications of Nanoscience and Nanotechnology: NSET Workshop Report. National Science Foundation. Available: http://itri.loyola.edu/nano/societalimpact/nanosi.pdf
6. Nanoscale Science, Engineering and Technology Subcommittee (NSET). 2006, July. The National Nanotechnology Initiative: Research and development leading to a revolution in technology and industry: Supplement to the Presidents 2007 budget. Available at: http://www.nano.gov/NNI_07Budget.pdf
7. Milburn, C. 2002. "Nanotechnology in the age of post human engineering: Science fiction as science." *Configurations* 10:261-295.
8. Keiper, A. 2003, Summer. "The nanotechology revolution." *The New Atlantis* 17-34.
9. Padeletti, G. & P. Fermo. 2003. "How the masters in Umbria, Italy, generated and used nanoparticles in art fabrication during the Renaissance period." *Applied Physics A: Materials Science & Processing* 76 (4):515-525.
10. Berube, D. M. 2006. *Nano-Hype: The truth behind the nanotechnology buzz*. New York: Prometheus Books.
11. National Nanotechnology Initiative (NNI). n.d. What is Nanotechnology? Retrieved May 14, 2007 at http://www.nano.gov/html/facts/whatIsNano.html
12. National Nanotechnology Initiative (NNI). n.d. About the NNI. Retrieved February 10, 2007 at http://www.nano.gov/html/about/home_about.html
13. National Nanotechnology Initiative (NNI). n.d. Structures and Strategies. Retrieved May 13, 2007 at http://www.nano.gov/html/about/nnistructure.html
14. President's Council of Advisors on Science and Technology. 2005. The National Nanotechnology Initiative at Five Years: Assessment and Recommendations of the National Nanotechnology Advisory Panel. Available at http://www.nano.gov//FINAL_PCAST_NANO_REPORT.pdf

15. National Nanotechnology Initiative (NNI). n.d. National Nanotechnology Initiative: FY 2008 Budget and Highlights. Retrieved May 14, 2007 at http://www.nano.gov/NNI_FY08_budget_summary-highlights.pdf

16. Roco, M.C. & W.C. Bainbridge, eds. 2005. Nanotechnology: Societal implications—Maximizing benefits for humanity. Report of the NNI Workshop: December 2-3, 2003. Available at: http://www.nano.gov/nni_societal_implications.pdf

17. Lane, N. & T. Kalil. 2005, Summer. "The National Nanotechnology Initiative: Present at the creation." Issues in Science and Technology Online. Retrieved May 10, 2007 at http://www.issues.org/issues/21.4/lane.html

18. Kurzweil, R. 2005. The singularity is near: When humans transcend biology. New York: Viking.

19. Kurzweil, R. 1999. *The age of spiritual machines: When computers exceed human intelligence.* New York: Penguin.

20. Senge, P. 1990. *The fifth discipline: The art and practice of the learning organization.* New York: Doubleday.

21. Mulhall, D. 2002. *Our molecular future: How nanotechnology, robotics, genetics, and artificial intelligence will transform our world.* Amherst, New York: Prometheus Books.

22. Courtney, H. 2002. 20/20 *foresight: Crafting strategy in an uncertain world.* Harvard Business School.

23. Mehta, M. and G. Hunt. 2006. "What makes nanotechnology special?" In G. Hunt and M. Mehta, eds. *Nanotechnology: Risk, Ethics and Law.* London: Earthscan.

CHAPTER 2

1. Oppenheimer, J. R. January 1946. Atomic Weapons. *Proceedings of the American Philosophical Society* 90(1):7-10.

2. Gaskell, G., E. Einsiedel, W. Hallman, S. H. Priest, J. Jackson, and J. Olsthoorn. 2005. "Social values and the governance of science." *Science,* 310 (December 23): 1908-1909.

3. Creswell, J. W. 1994. *Research design: Qualitative and quantitative approaches.* Thousand Oaks: Sage.

4. Chadwick, R. 2005. "Professional ethics and the 'good' of science." *Interdisciplinary Science Reviews* 30(3):247-256.

5. Bentz, V. M. and J.J. Shapiro. 1998. *Mindful inquiry in social research.* Thousand Oaks: Sage.

6. Harding, S. 2002. "Must the advance of science advance global inequality?" *International Studies Review* 4(2):87-105.

7. Allen, T.F.H., J. A. Tainter, J. C. Pires, and T. W. Hoekstra. 2001. "Dragnet ecology—'Just the facts, Ma'm': The privilege of science in a postmodern world." *BioScience* 51(6):475-485.

8. Jasanoff, S. 2004. *States of Knowledge: The Co-Production of Science and Social Order.* London: Routledge.

9. Wynne, B. 2001. "Creating public alienation: Expert cultures of risk and ethics on GMOs." *Science as Culture* 10(4):445-481.

10. See, for example, the various codes of ethics available through The Online Ethics Center for Engineering and Science at Case Western Reserve University. Available at: http://onlineethics.org/codes/index.html.

11. Levitt, N. and P. R. Gross. 1996. "Academic anti-science." *Academe* (November-December): 38-42.
12. Jonas, H. 2004/1979. "Toward a philosophy of technology." In D. M. Kaplan, ed. *Readings in the Philosophy of Technology* 17-33. New York: Rowman & Littlefield.
13. Mehta, M. D. and G. Hunt. 2006. "What makes nanotechnologies special?" In G. Hunt and M. Mehta, eds. *Nanotechnology: Risk, ethics and law* 273-281. London: Earthscan.
14. Emison, G. A. 2004. "American pragmatism as a guide for professional ethical conduct for engineers." *Science and Engineering Ethics* 10(2):225-233.
15. Uff, J. n.d. *Engineering ethics: Do engineers owe duties to the public?* London: Royal Academy of Engineering. Retrieved June 25, 2007 at http://www.raeng.org.uk/news/publications/list/lectures/Engineering_Ethics_Lecture.pdf
16. d'Anjou, P. 2004. "Theoretical and methodological elements for integrating ethics as a foundation into the education of professional and design disciplines." *Science and Engineering Ethics* 10(2):211-218.
17. Bird, S. J. 2002. "The processes of science." In R. Spier, ed. *Science and technology ethics*. London: Routledge.
18. Johnson, C. 2007. *Ethics in the workplace: Tools and tactics for organizational transformation*. Thousand Oaks: Sage.
19. Bainbridge, W. 2002. "Public attitudes toward nanotechnology." *Journal of Nanoparticle Research* 4(6):461-470.
20. Cobb, M. and J. Macoubrie. 2004. "Public perception about nanotechnology: Risks, benefits and trust." *Journal of Nanoparticle Research* 6(4):395-405.
21. Siegrist, M., C. Keller, H. Kastenholz, S. Frey, and A. Wiek. 2007. "Laypeople's and experts perception of nanotechnology hazards." *Risk Analysis* 27(1):59-69.
22. Cobb, M. 2005. "Framing effects on public opinion about nanotechnology." *Science Communication* 27(2):221-239.
23. Balbus, J., R. Denison, K. Florini, and S. Walsh. 2006. "Getting nanotechnology right the first time." In G. Hunt and M. Mehta, eds. *Nanotechnology: Risk, Ethics and Law*. London: Earthscan.
24. Macoubrie, J. 2006. "Nanotechnology: Public concerns, reasoning and trust in government." *Public Understanding of Science* 15(2):221-241.
25. Fisher, E. and R. L. Mahajan. 2006. "Contradictory intent? U.S. federal legislation on integrating societal concerns into nanotechnology research and development." *Science and Public Policy* 33(1):5-16.
26. Berube, D. M. 2006. *Nano-hype: The truth behind the nanotechnology buzz*. New York: Prometheus Books.
27. Priest, S. H. 2005. "Commentary—Room at the bottom of Pandora's box: Peril and promise in communicating technology." *Science Communications* 27(2):292-299.

CHAPTER 3

1. Roco, M. C. and W. S. Bainbridge, eds. 2001. *Societal implications of nanoscience and nanotechnology*. Dordrecht: Kluwer Academic Publishers. Available online at: http://www.wtec.org/loyola/nano/societalimpact/nanosi.pdf
2. Rifkin, J. 1999. *The biotech century: Harnessing the gene and remaking the world*. New York: Tarcher/Putnam.

3. Bond, P. J. 2005. "Preparing the path for nanotechnology." In M. C. Roco and W. C. Bainbridge, eds. *Nanotechnology: Societal implications—Maximizing benefits for humanity*. Report of the NNI Workshop: December 2-3, 2003 (16-21). Available at: http://www.nano.gov/nni_societal_implications.pdf

4. Schummer, J. 2005. "Societal and ethical implications of nanotechnology: Meanings, interest groups and social dynamics." *Techne: Research in Philosophy and Technology* 8(2). Retrieved June 6, 2007 at http://scholar.lib.vt.edu/ejournals/SPT/v8n2/schummer.html

5. Creighton, M. 2002. *Prey*. New York: HarperCollins.

6. Berne, R. W. and J. Schummer. 2005. "Teaching societal and ethical implications of nanotechnology to engineering students through science fiction." *Bulletin of Science, Technology and Society* 25(6):459-468.

7. Kurzweil, R. 2005. *The singularity is near: When humans transcend biology*. New York: Viking.

8. Berube, D. 2006. *Nano-Hype: The truth behind the nanotechnology buzz*. New York: Prometheus.

9. Mills, K. and C. Fleddermann. 2005. "Getting the best from nanotechnology: Approaching social and ethical implications openly and proactively." *IEEE Technology and Society Magazine* Winter: 18-26.

10. Lewenstein, B. V. 2005. "What counts as a 'societal and ethical issue' in nanotechnology?" *International Journal for Philosophy of Chemistry* 11(1):5-18.

11. Gaskell, G., E. Einsiedel, W. Hallman, S. H. Priest, J. Jackson, and J. Olsthoorn. 2005. "Social values and the governance of science." *Science* 310 (December 23): 1908-1909.

12. Kulinowsky, K. 2006. "Nanotechnology: From 'Wow' to 'Yuck'?" In G. Hunt and M. Mehta, eds. *Nanotechnology: Risk, ethics and law* (14-24). London: Earthscan.

13. Mehta, M. D. and G. Hunt. 2006. "What makes nanotechnologies special?" In G. Hunt and M. Mehta, eds. *Nanotechnology: Risk, ethics and law* (273-281). London: Earthscan.

14. See the website for the Human Genome Project for more detailed information. Available at: http://www.ornl.gov/sci/techresources/Human_Genome/home.shtml

15. Human Genome Project. n.d. *Ethical, legal, and social issues*. Available at: http://www.ornl.gov/sci/techresources/Human_Genome/elsi/elsi.shtml

16. ELSI Research Planning and Evaluation Group. 2000, February. A review and analysis of the ethical, legal and social implications (ELSI) research programs at the National Institutes of Health and the Department of Energy. Retrieved April 21, 2007 at http://www.genome.gov/Pages/Hyperion/About_NHGRI/Der/Elsi/erpeg_report.pdf

17. Fisher, E. 2005. "Lessons learned from the ethical, legal, and social implications program (ELSI): Planning societal implications research for the National Nanotechnology Program." *Technology in Society* 27:321-328.

18. Ramsay, S. 2001. "Ethical implications of research on the human genome." *The Lancet* 357 (February 17): 535.

19. President's Council of Advisors on Science and Technology. 2005. *The National Nanotechnology Initiative at five years: Assessment and recommendations of the National Nanotechnology Advisory Panel*. Available at: http://www.nano.gov//FINAL_PCAST_NANO_REPORT.pdf

20. Nanoscale Science, Engineering and Technology Subcommittee (NSET). 2004. National Nanotechnology Initiative. 2004. *The National Nanotechnology Initiative Strategic Plan*. National Science and Technology Council. Available at: http://www.nano.gov/NNI_Strategic_Plan_2004.pdf

21. Nanoscale Science, Engineering and Technology Subcommittee (NSET). 2006, July. *The National Nanotechnology Initiative: Research and development leading to a revolution in technology and industry: Supplement to the Presidents 2007 budget*. Available at: http://www.nano.gov/NNI_07Budget.pdf

22. National Nanotechnology Initiative. n.d. *FY 2008 Budget and Highlights*. Retrieved June 30, 2007 at: http://www.nano.gov/NNI_FY08_budget_summary-highlights.pdf

23. Sandler, R. and W. D. Kay. 2006. "The National Nanotechnology Initiative and the social good." *The Journal of Law, Medicine, and Ethics* 34(4):675-681.

24. World Commission on the Ethics of Scientific Knowledge and Technology (COMEST). 2005, March. *The Precautionary Principle*. Paris: United Nations Educational, Scientific and Cultural Organization. Retrieved April 10, 2007 at http://unesdoc.unesco.org/images/0013/001395/139578e.pdf

25. Gardiner, S. M. 2006. "A core precautionary principle." *The Journal of Political Philosophy* 14(1):33-60.

26. Resnick, D. B. 2003. "Is the precautionary principle unscientific?" *Studies in History and Philosophy of Biological and Biomedical Sciences* 34:329-344.

27. Ascher, W. 2004. *Scientific information and uncertainty: Challenges for the use of science in policymaking* 10:437-455.

28. Harremoes, P., D. Gee, M. MacGarvin, A. Stirling, J. Keys, B. Wynne, and S. G. Vaz. 2001. *Late lessons from early warnings: the precautionary principle 1896-2000*. European Environment Agency. Retrieved May 1, 2007 at http://reports.eea.europa.eu/environmental_issue_report_2001_22/en

29. National Science Board. 2006. *Science and Engineering Indicators 2006*. Two volumes. Arlington, VA: National Science Foundation (volume 1, NSB 06-01; volume 2, NSB 06-01A).

CHAPTER 4

1. United Nations. 1948. *Universal Declaration of Human Rights*. Retrieved May 28, 2007 at http://www.un.org/Overview/rights.html

2. For a more detailed description of ethical theories and principles, as well as links to additional resources, see: Bennett-Woods, D. 2005. *Ethics at a glance*. Regis University online reference. Available at: http://support.regis.edu/shcp/shcp_forum/Interactions/EthicsAtAGlance/index.html

3. For information on the Tuskegee Study, see: Centers for Disease Control and Prevention n.d.. *U.S. Public Health Service Study of Syphilis at Tuskegee*. National Institutes of Health website. Available at: http://www.cdc.gov/nchstp/od/tuskegee/time.htm

4. Bennett-Woods, D. and E. Fisher. 2004. *Nanotechnology and the IRB: Toward a New Paradigm for Analysis and Dialogue*. Paper presented in the session entitled Nanotechnology: Risk, Rhetoric and Imagination. Joint meeting of the European Association for the Study of Science and Technology, and Society for Social Studies of Science. Paris, France. 2004. *Conference theme: Public Proofs: Science Technology and Democracy*. Paper available online: http://www.csi.ensmp.fr/csi/4S/index.php

5. Sarewitz D. and E. Woodhouse. 2003. "Small is powerful." In A. Lightman, D. Sarewitz, and C. Desser, eds. *Living with the Genie: Essays on Technology and the Quest for Human Mastery*, pp. 63-83. Washington: Island Press.
6. "The National Commission for the Protection of Human Subjects of Biomedical and Behavioral Research." *The Belmont Report*. 1979. Available online at http://ohsr.od.nih.gov/guidelines/belmont.html
7. Bennett-Woods, D. 2007. "Integrating ethical considerations into funding decisions for emerging technologies." *Journal of Nanotechnology Law and Business* 4(1), online.
8. Grunwald, A. 2005. "Nanotechnology—A new field of ethical inquiry?" *Science and Engineering Ethics* 11(2):187-201.
9. Lewenstein, B.V. 2005. "What counts as a 'social and ethical issue' in nanotechnology?" *HYLE—International Journal for Philosophy of Chemistry* 11(1):5-18.
10. Bennett-Woods, D. 2006. "Nanotechnology in Medicine: Implications of Converging Technologies on the Human Community." *Development* 49(4):54-59.
11. Foresight Nanotech Institute. *Foresight Guidelines for Responsible Nanotechnology Development*, at http://www.foresight.org/guidelines/current.html (last visited May 29, 2006). (This is the sixth draft version of the guidelines; there have been five previous versions.)

CHAPTER 5

1. *American Heritage Dictionary of the English Language* (4th edition). 2000. Houghton Mifflin Company. Online version at: http://www.thefreedictionary.com/discernment
2. Johnson, C. E. 2005. *Meeting the ethical challenges of leadership: Casting light or shadow* (2nd ed.). Thousand Oaks, CA: Sage.
3. Bennett-Woods, D. 2007. Integrating ethical considerations into funding decisions for emerging technologies. *Journal of Nanotechnology Law and Business* 4(1), online.
4. Berube, D. M. 2006. *Nano-Hype: The truth behind the nanotechnology buzz.* New York: Prometheus Books.
5. Roco, M.C. and W. C. Bainbridge, eds. 2001. *Societal Implications of Nanoscience and Nanotechnology*: NSET Workshop Report. National Science Foundation. Available: http://itri.loyola.edu/nano/societalimpact/nanosi.pdf
6. Bennett-Woods, D. and E. Fisher. 2004. *Nanotechnology and the IRB: Toward a new paradigm for analysis and dialogue.* Joint meeting of the European Association for the Study of Science and Technology and Society for Social Studies of Science. Paris, France. 2004. Conference theme: Public Proofs: Science Technology and Democracy. Paper available online: http://www.csi.ensmp.fr/csi/4S/index.php
7. Fisher, E. and R. L. Mahajan. 2006. *Midstream modulation of nanotechnology research in an academic laboratory.* Proceedings of the ASMA International Mechanical Engineering Congress and Exposition (November 2006), Chicago.
8. Ihde, D. 1993. *Philosophy of Technology: An Introduction.* New York: Paragon.
9. The Royal Society and the Royal Academy of Engineering. 2004. *Nanoscience and Nanotechnologies: Opportunities and Uncertainties.* London: The Royal Society. Available online at: http://www.nanotec.org.uk/finalReport.htm
10. Tepper, A. 1996. "Controlling technology by shaping visions." *Policy Sciences* 29(1):29-44.

11. Pielke, R.A., Jr., R. Byerly, Jr. 1998. "Beyond basic and applied." *Physics Today* 51(2):42-46.
12. Bush, V. 1948. *Science: The endless frontier*. Washington: United States Government Printing Office.
13. Branscomb, L. and R. Florida. 1998. "Challenges to technology policy in a changing world economy." In L. Branscomb and J. Keller, eds. *Investing in innovation: Creating a research and innovation policy that works* (pp. 3-39). Cambridge: MIT Press.
14. Harremoes, P., D. Gee, M. MacGavin, A. Stirling, J. Keys, B. Wynne, and S. Vaz, eds. 2001. *Late lessons from early warnings: The precautionary principle 1896-2000*. European Environment Agency. Retrieved February 23, 2006 at http://reports.eea.europa.eu/environmental_issue_report_2001_22/en

CHAPTER 6

1. National Science and Technology Council. 1999. *Nanotechnology: Shaping the world atom by atom*. Available at: http://www.wtec.org/loyola/nano/IWGN. Public.Brochure/IWGN.Nanotechnology.Brochure.pdf
2. Roco, M. C. 2004. "Nanoscale science and engineering: Unifying and transforming tools." *American Institute of Chemical Engineers (AIChE) Journal* 50(5):890-897.
3. Joy, B. 2000. "Why the future doesn't need us." *Wired* 8(4): online. Retrieved April 26, 2004 at http://www.wired.com/wired/archive/8.04/joy.html
4. Berger, M. April 2007. "Debunking the trillion dollar nanotechnology market size hype." *Nanowerk*. Retrieved September 1, 2007 at http://www.nanowerk. com/spotlight/spotid=1792.php
5. Technology Transfer Center as cited in D. Stark. June 2007. *Nanotechnology in Europe—Ensuring the EU Competes Effectively on the World Stage: Survey & Workshop*. Nanoforum. Düsseldorf, Germany.
6. Gleiche, M., H. Hoffschulz, and S. Lenhert. October 2006. "Nanoforum report: Nanotechnology in consumer products." *Nanoforum.org*, European Nanotechnology Gateway. Available at http://www.nanoforum.org/dateien/temp/ Nanoforum_NanotechCommercialisation_Final.pdf?02092007064205
7. Roco, M. C. and W. C. Bainbridge, eds. 2001. *Societal Implications of Nanoscience and Nanotechnology: NSET Workshop Report*. National Science Foundation. Available: http://itri.loyola.edu/nano/societalimpact/nanosi.pdf
8. Weil, V. 2001. "Ethical issues in nanotechnology." In M. C. Roco and W. C. Bainbridge, eds., 193-198. *Societal Implications of Nanoscience and Nanotechnology: NSET Workshop Report*. National Science Foundation. Available: http://itri. loyola.edu/nano/societalimpact/nanosi.pdf
9. Mehta, M. D. and G. Hunt. 2006. "What makes nanotechnologies special?" In G. Hunt and M. Mehta, eds. *Nanotechnology: Risk, Ethics, and Law* (273-281). London: Earthscan.
10. Bennett-Woods, D. 2007. "Integrating ethical considerations into funding decisions for emerging technologies." *Journal of Nanotechnology Law and Business* 4(1), online.
11. Bennett-Woods, D. 2006. "Nanotechnology in medicine: Implications of converging technologies on the human community." *Development* 49(4):54-59.

12. Bennett-Woods, D. and E. Fisher. 2004. *Nanotechnology and the IRB: Toward a new paradigm for analysis and dialogue.* Paper presented in the session entitled Nano-technology: Risk, Rhetoric, and Imagination. Joint meeting of the European Association for the Study of Science and Technology, and Society for Social Studies of Science. Paris, France. 2004. Conference theme: Public Proofs: Science Technology and Democracy. Paper available online: http://www.csi.ensmp.fr/csi/4S/index.php

13. Sarewitz, D. 2003. "Science and happiness." In A. Lightman, D. Sarewitz and C. Desser, eds. *Living with the Genie: Essays on Technology and the Quest for Human Mastery* (181-200). Washington: Island Press.

14. Visvanathan, S. 2003. "Progress and violence." In A. Lightman, D. Sarewitz and C. Desser, eds. *Living with the Genie: Essays on Technology and the Quest for Human Mastery* (157-180). Washington: Island Press.

15. Lightman, A., D. Sarewitz, and C. Desser. 2003. "Introduction." In A. Lightman, D. Sarewitz and C. Desser, eds. *Living with the Genie: Essays on Technology and the Quest for Human Mastery* (1-4). Washington: Island Press.

16. Etzioni, A. 1995. *Rights and the Common Good: The Communitarian Perspective.* New York: St. Martin's Press.

17. Reeve, A. 2002. "A social contract?" In R. E. Spier, ed. *Science and Technology Ethics* (107-126). London: Routledge.

18. Harremoes, P., D. Gee, M. MacGarvin, A. Stirling, J. Keys, B. Wynne, and S. G. Vaz. 2001. *Late lessons from early warnings: the precautionary principle 1896-2000.* European Environment Agency. Retrieved May 1, 2007 at http://reports.eea.europa.eu/environmental_issue_report_2001_22/en

19. See publications on the National Nanotechnology Initiative (NNI) website at http://www.nano.gov/index.html

20. Hett, A. 2007. "Nanotechnology and the two faces of risk from a reinsurance perspective." In N. M. de S. Cameron and M. Ellen Mitchell, eds. *Nanoscale: Issues and Perspectives for the Nano Century* (15-26). Hoboken, NJ: John Wiley & Sons.

21. National Institute for Occupational Safety and Health (NIOSH). (February 2007). Progress toward safe nanotechnology in the workplace. Department of Health and Human Services, Washington, DC. Available online at: http://www.cdc.gov/niosh/docs/2007-123/pdfs/2007-123.pdf

22. Smith, R. H. 2001. "Social, ethical and legal implications of nanotechnology." In M. C. Roco and W. S. Bainbridge, eds. *Societal Implications of Nanoscience and Nanotechnology: NSET Workshop Report* (pp. 203-210). National Science Foundation. Available: http://itri.loyola.edu/nano/societalimpact/nanosi.pdf

CHAPTER 7

1. Rowling, J. K. 2005. *Harry Potter and the Half-Blood Prince.* New York: Arthur A. Levine Books.

2. Roco, M.C. 2006. "National Nanotechnology Investment in the FY 2007 Budget Request." In the Intersociety Working Group of the American Association for the Advancement of Science, *AAAS Report XXXI: Research and Development: FY 2007.* Retrieved online April 11, 2007 at http://www.aaas.org/spp/rd/07pch24.htm

3. National Nanotechnology Initiative 2007. *National Nanotechnology Initiative: FY 2008 Budget and Highlights.* Retrieved July 21, 2007 at http://www.nano.gov/ NNI_FY08_budget_summary-highlights.pdf

4. Altmann, J. 2006. *Military Nanotechnology: Potential Applications and preventive arms control.* New York: Routledge.

5. Shipbaugh, C. 2006. "Offense-defense aspects of nanotechnologies: A forecast of potential military applications." *Journal of Law, Medicine, and Ethics* 34(4):741-747.

6. Ratner, D. and M. A. Ratner. 2004. *Nanotechnology and Homeland Security: New Weapons for New Wars.* Upper Saddle River, NJ: Prentice Hall.

7. Cobb, M.D. and J. Macoubrie. 2004. "Public perceptions about nanotechnology: Risks, benefits, and trust." *Journal of Nanoparticle Research* 6:395-405.

8. All references to the MIT Institute for Soldier Nanotechnologies cite information available on the Institute's website at http://web.mit.edu/ISN/ as of August 2007.

9. All references to the Future Force Warrior Program of the U.S. Army Natick Soldier Research, Development, and Engineering Center cite information available on the Center's website at http://www.natick.army.mil/soldier/WSIT/ as of August 2007.

10. All references to the Defense Advanced Research Projects Agency (DARPA) cite information available on the department's website at http://www.darpa.mil/

11. Bennett-Woods, D. and E. Fisher. 2004. *Nanotechnology and the IRB: Toward a New Paradigm for Analysis and Dialogue.* Paper presented in the session entitled Nanotechnology: Risk, Rhetoric, and Imagination. Joint meeting of the European Association for the Study of Science and Technology and Society for Social Studies of Science. Paris, France. Conference theme: Public Proofs: Science Technology and Democracy. Paper available online: http://www.csi.ensmp.fr/csi/4S/index.php

12. Ricks, T. E. August 29, 2007. "Bush wants $50 billion more for Iraq War." *Washington Post* A01.

13. Bilmes, L. and J. Stiglitz. 2006. *The economic costs of the Iraq war: An appraisal three years after the beginning of the conflict.* National Bureau of Economic Research Working Paper. Retrieved August 15, 2007 at http://works.bepress.com/cgi/ viewcontent.cgi?article=1009&context=joseph_stiglitz

14. Hunt, G. 2006. "The Global Ethics of Nanotechnology." In G. Hunt and Michael M., eds. *Nanotechnology: Risk, ethics, and law.* London: Earthscan.

15. Mehta, M. 2002. "Privacy vs. surveillance: How to avoid a nano-panoptic future." *Canadian Chemical News* 54(10):31-33.

16. Flagg, B. N. 2005. *Nanotechnology and the public.* Part I of front-end analysis in support of Nanoscale Informal Science Education (NISE) Network. Available online at: http://www.informalscience.org/download/case_studies/report_ 149.pdf

17. Scheufele, D. A. and B. V. Lewenstein. 2005. "The public and nanotechnology: How citizens make sense of emerging technologies." *Journal of Nanoparticle Research* 7:659-667.

18. Cobb, M. and J. Macoubrie. 2004. "Public perception about nanotechnology: Risks, benefits, and trust." *Journal of Nanoparticle Research* 6(4):395-405.

19. Heller, J. and C. Peterson. 2007. "Nanotechnology: Maximizing benefits, minimizing downsides." In N. M. Cameron and M. E. Mitchell, eds. *Nanoscale: Issues and perspectives for the nano century.* Chicago: John Wiley & Sons.

20. See the following websites for example: FERPA at http://www.ed.gov/ policy/gen/guid/fpco/ferpa/index.html and HIPAA at http://www.hhs. gov/ocr/hipaa/
21. For more information refer to the Electronic Privacy Information Center at http://www.epic.org/privacy/terrorism/fisa/

CHAPTER 8

1. Carson, R. L. 1962. *Silent spring* (40th Anniversary Edition). New York: First Mariner Books.
2. See http://www.amazon.com/Genres-Science-Fiction-Environmental-Catas-trophe/lm/35B174398J05Y. List created by C. R. Oseland.
3. Schumacher, E. F. 1989. *Small is beautiful: Economics as if people mattered* (re-issue of first edition, 1973). New York: Harper & Row.
4. Smith, D. S. 2001. "Place-based environmentalism and global warming: Conceptual contradictions of American environmentalism." *Ethics and International Affairs* 15(2):117-135.
5. Lester, J. P. 1998. "Looking backward to see ahead: The evolution of environmental politics and policy, 1890-1998." *Forum for Applied Research and Public Policy* 13(4):30-36.
6. Agyeman, J., R. D. Bullard, and B. Evans. 2003. "Joined-up thinking: Bringing together sustainability, environmental justice and equity." Introduction to J. Agyeman, R. D. Bullard and B. Evans, eds. *Just sustainabilities: Development in an unequal world* 1-16. Cambridge, MA: MIT Press.
7. World Commission on Environment and Development. 1987. *Our common future.* Available at: http://www.unngocsd.org/documents/brundtland_bericht.pdf
8. McLaren, D. 2003. "Environmental space, equity and the ecological debt." In J. Agyeman, R. D. Bullard and B. Evans, eds. *Just sustainabilities: Development in an unequal world* 19-37. Cambridge, MA: MIT Press.
9. Faber, D. R. and D. McCarthy. 2003. "Neo-liberalism, globalization, and the struggle for economic democracy: Linking sustainability and environmental justice." In J. Agyeman, R. D. Bullard and B. Evans, eds. *Just sustainabilities: Development in an unequal world* 38-63. Cambridge, MA: MIT Press.
10. Commission on the Ethics of Scientific Knowledge and Technology. March 2005. *The precautionary principle.* United Nations Educational, Scientific, and Cultural Organization, Paris. Available at: unesdoc.unesco.org/images/0013/001395/139578e.pdf
11. Kimbrell, G. A. 2007. "The potential environmental hazards of nanotechnology and the applicability of existing law." In N. M. de S. Cameron and M. E. Mitchell, eds. *Nanoscale: Issues and perspectives for the nano century* 211-238. Hoboken, NJ: John Wiley & Sons.
12. Clift, R. 2006. "Risk Management and Regulation in an Emerging Technology." In G. Hunt and M. Mehta, eds. *Nanotechnology: Risk, ethics and law* 140-153. London: Earthscan.
13. Colvin, V. L. 2003. "The potential environmental impact of engineered nano-materials." *Nature Biotechnology* 21(10):1166-1170.
14. Sweet, L. and B. Strohm. 2006. "Nanotechnology: Life-cycle risk management." *Human and Ecological Risk Assessment* 12:528-551.

15. Blackwelder, B. 2007. "Nanotechnology jumps the gun: Nanoparticles in consumer products." In N. M. de S. Cameron and M. E. Mitchell, eds. *Nanoscale: Issues and perspectives for the nano century* 71-82. Hoboken, NJ: John Wiley & Sons.

16. Uskokovic, V. 2007. "Nanotechnologies: What we do not know." *Technology in Society* 29:43-61.

17. Singer, P.A, F. Salamanca-Buentello, and A. S. Darr. Summer 2005. "Harnessing nanotechnology to improve global equity." *Issues in Science and Technology Online* 57-64.

18. Meridian Institute. January 2005. *Nanotechnology and the poor: Opportunities and risks*. Meridian Institute, Dillon, CO. Available at http://www/nanoandthepoor.org

19. Hett, A. 2007. "Nanotechnology and the two faces of risk from a reinsurance perspective." In N. M. de S. Cameron and M. E. Mitchell, eds. *Nanoscale: Issues and perspectives for the nano century* 15-26. Hoboken, NJ: John Wiley & Sons.

20. Resnick, D. B. 2002. "Is the precautionary principle unscientific?" *Studies in History and Philosophy of Biological and Biomedical Sciences* 34:329-344.

21. Hahm, R. W. and C. R. Sunstein. 2005. "The precautionary principle as a basis for decision making." *The Economists' Voice* 2(2):e1-9.

22. Harremoes, P., D. Gee, M. MacGarvin, A. Stirling, J. Keys, B. Wynne, and S. G. Vaz. 2001. *Late lessons from early warnings: The precautionary principle 1896-2000*. European Environment Agency. Retrieved May 1, 2007 at http://reports.eea.europa.eu/environmental_issue_report_2001_22/en

23. Phoenix, C. and M. Treder. 2004. *Applying the precautionary principle to nanotechnology*. Center for Responsible Nanotechnology. Retrieved September 6, 2007 at http://www.crnano.org/precautionary.htm

24. Mehta, M. D. and G. Hunt. 2006. "What makes nanotechnologies special?" In G. Hunt and M. Mehta, eds. *Nanotechnology: Risk, Ethics, and Law* 273-281. London: Earthscan.

25. See the UN Millennium Development Goals at http://www.un.org/millenniumgoals/

26. Salamanca-Buentello, F., D. L. Persad, E. B. Court, D. K. Martin, A. S. Daar, and P. A. Singer. 2005. "Nanotechnology and the developing world." *PloS Med* 2(4): e97.

27. Daar, A. S., H. Thorsteinsdóttir, D. K. Martin, A. C. Smith, S. Nast, and P. A. Singer/ 2002. "Top Ten Biotechnologies For Improving Health in Developing Countries." *Nature Genetics* 32:229-232.

28. World Health Organization. 2002. *The World Health Report 2002: Reducing risks, promoting healthy life* (published report). Geneva: World Health Organization.

29. Bennett-Woods, D. 2006. "Nanotechnology in medicine: Implications of conversing technologies on the human community." *Development* 49(4):54-58.

30. Invernizzi, N. and G. Foladori. 2006. "Nanomedicine, poverty and development." *Development* 49(4):114-118.

31. Maynard, A. D. June 25, 2007. *Developing science policies for sustainable nanotechnologies*. Presentation to the President's Council of Advisors on Science and Technology. Available at www.nanotechproject.org/file_download/202

32. National Nanotechnology Initiative. September 2006. *Environmental, health, and safety research needs for engineered nanoscale materials*. Retrieved June 28, 2007 at http://www.nano.gov/NNI_EHS_research_needs.pdf

33. Nanotechnology Environmental and Health Implications Working Group. August 2007. *Prioritization of environmental, health and safety research needs for engineered nanoscale materials: An interim document for public comment.* National Science and Technology Council. Retrieved September 6, 2007 at http://www.nano.gov/Prioritization_EHS_Research_Needs_Engineered_Nanoscale_Materials.pdf

34. As cited in Gaither, C. C. and A. E. Cavazos-Gaither. 1998. *Practically speaking: A dictionary of quotations on engineering, technology and architecture.* Bristol: Institute of Physics Publishing.

35. Phelps, T. A. 2007. "The European approach to nanoregulation." In N. M. de S. Cameron and M. E. Mitchell, eds. *Nanoscale: Issues and perspectives for the nano century* 189-210. Hoboken, NJ: John Wiley & Sons.

36. Hunt, G. 2006. "Nanotechnologies and society in Europe." In G. Hunt and M. Mehta, eds. *Nanotechnology: Risk, Ethics, and Law* 92-104. London: Earthscan.

37. Goldenburg, L. 2006. "Nanotechnologies and society in Canada." In G. Hunt and M. Mehta, eds. *Nanotechnology: Risk, Ethics, and Law* 105-117. London: Earthscan.

38. Masami, M., G. Hunt, and O. Masayuki. 2006. "Nanotechnologies and society in Japan." In G. Hunt and M. Mehta, eds. *Nanotechnology: Risk, Ethics, and Law* 59-73. London: Earthscan.

39. Mills, K. 2006. "Nanotechnologies and society in the USA." In G. Hunt and M. Mehta, eds. *Nanotechnology: Risk, Ethics, and Law* 74-91. London: Earthscan.

40. Balbus, J., R. Denison, K. Florini, and S. Walsh. Summer 2005. "Getting nanotechnology right the first time." *Issues in Science and Technology Online.* Available at http://www.issues.org/21.4/balbus.html

41. For an interesting discussion of the intersection of technology, economics, and human nature, see Small, B. and N. Jollands. 2006. "Technology and ecological economics: Promethean technology, Pandorian potential." *Ecological economics* 56:343-358.

CHAPTER 9

1. Amiel, H.F. as cited on *GIGA Quotes* at http://www.giga-usa.com/quotes/authors/henri-frederic_amiel_a001.htm

2. Bennett-Woods, D. 2007. "Anticipating the impact of nanoscience and nanotechnology in health care." In N. M. Cameron and M. E. Mitchell, eds. *Nanoscale: Issues and perspectives for the nano century.* Hoboken, N.J.: John Wiley & Sons.

3. Macoubrie, J. 2006. "Nanotechnology: Public concerns, reasoning and trust in government." *Public Understanding of Science* 15(2):221-241.

4. Cobb, M. and J. Macoubrie. 2004. "Public perception about nanotechnology: Risks, benefits and trust." *Journal of Nanoparticle Research* 6(4):395-405.

5. Vo-Dinh, T. 2007. "Nanotechnology in biology and medicine: The new frontier." In T. Vo-Dinh, ed. *Nanotechnology in biology and medicine: Methods, devices and applications* 1.1-1.9. Boca Raton, FL: CRC Press.

6. Chan, W. C. W. 2006. "Bionanotechnology progress and advances." *Biology of Blood and Marrow Transplantation* 12:87-91.

7. Ebbesen, M. and T. G. Jensen. 2006. "Nanomedicine: Techniques, potentials, and ethical implications." *Journal of Biomedicine and Biotechnology* 2006:1-11.

8. Roco, M. C. 2005. "Converging technologies: Nanotechnology and biomedicine." In N. H. Malsch, ed. *Biomedical nanotechnology* pp. xi-ixx. Boca Raton, FL: Taylor & Francis.

9. European Technology Platform NanoMedicine. 2006, November. *Nanomedicine: Nanotechnology for health: Strategic research agenda for nanomedicine.* European Communities. Luxembourg: Office for Official Publications of the European Communities. Available at: http://cordis.europa.eu/nanotechnology/nanomedicine.htm#terms

10. National Nanotechnology Initiative. 2007. *National Nanotechnology Initiative: FY 2008 budget and highlights.* Available at: http://www.nano.gov/NNI_FY08_budget_summary-highlights.pdf

11. Marcus Tullius Cicero as cited on *GIGA quotes* at http://www.giga-usa.com/quotes/authors/cicero_a001.htm

12. Ferrari, M. and G. Downing. 2005. "Medical nanotechnology: Shortening clinical trials and regulatory pathways?" *Biodrugs* 19(4):203-210.

13. Singer, E., K. Lamparska-Kupsik, J. Clark, K. Munson, L. Kretzner, and S. S. Smith. 2007. "Nucleoprotein-based nanodevices in drug design and delivery." In T. Vo-Dinh, ed. *Nanotechnology in biology and medicine: Methods, devices, and applications* 7-1--7-15. Boca Raton, FL: CRC Press.

14. Alonso, M. J. 2004. "Nanomedicines for overcoming biological barriers." *Biomedicine and Pharmacotherapy* 58(3):168-172.

15. Kubik, T., K. Bogunia-Kubik, and M. Sugisaka 2005. "Nanotechnology on duty in medical applications." *Current Pharmaceutical Biotechnology* 6:17-33.

16. Bullis, K. (March/April 2006). "Nanomedicine." *Technology Review* 58-59.

17. Kayser, O., A. Lemke, and N. Hernández-Trejo. 2005. "The impact of nanobiotechnology on the development of new drug delivery systems." *Current Pharmaceutical Biotechnology* 6:3-5.

18. Lindpainter, K. 2003. "Pharmacogenetics and the future of medical practice." *Journal of Molecular Medicine* 81:141-153.

19. Vo-Dinh, T. 2007b. "Optical nanobiosensors and nanoprobes." In T. Vo-Dinh, ed. *Nanotechnology in biology and medicine: Methods, devices, and applications* 17-1--17-10. Boca Raton, FL: CRC Press.

20. Choi, Y. and J. R. Baker. 2007. "Nanoparticles in medical diagnostics and therapeutics." In T. Vo-Dinh, ed. *Nanotechnology in biology and medicine: Methods, devices and applications* 31-1--31-22. Boca Raton, FL: CRC Press.

21. Mazzola, L. 2003. "Commercializing Nanotechnology." *Nature Biotechnology* 21(10):1137-1143.

22. Campo, A. and I. J. Bruce. 2005. "Diagnostics and High Throughput Screening." In N. H. Malsch, ed. *Biomedical Nanotechnology.* Boca Raton: Taylor & Francis.

23. Van den Bueken, X., F. Walboomers, and J. A. Jansen. 2005. "Implants and prostheses." In N. H. Malsch, ed. *Biomedical Nanotechnology.* Boca Raton: Taylor & Francis.

24. Chang, T. M. S. 2005. "Therapeutic applications of polymeric artificial cells." *Nature Reviews* 4:221-235.

25. Patel, G.M., G. C. Patel, R. B. Patel, J. K. Patel, and M. Patel. 2006. "Nanorobot: A versatile tool in nanomedicine." *Journal of Drug Targeting* 14(2):63-67.

26. Crichton, M. 2002. *Prey.* New York: Avon Books.

27. Weston, A. D. and L. Hood. 2004. "Systems biology, proteomics, and the future of health care: Toward predictive, preventative, and personalized medicine," *Journal of Proteome Research* 3:179-196.

28. Califf, R. M. 2004. "Defining the balance of risk and benefit in the era of genomics and proteomics." *Health Affairs* 23(1):77-87.

29. Pison, U., T. Welte, M. Giersig, and D. A. Groneberg. 2006. "Nanomedicine for respiratory diseases." *European Journal of Pharmacology* 533:341-350.

30. Betta, M. and V. Clulow. Spring 2005. "Healthcare management: Training and education in the genomic era." *Journal of Health and Human Services Administration*, 465-500.

31. National Center for Health Statistics 2006. *Health, United States, 2006.* U.S. Department of Health and Human Services. Hyattsville, MD. http://www.cdc.gov/nchs/data/hus/hus06.pdf#executivesummary

32. U.S. Bureau of Census. 2007, August. *Income, poverty and health insurance coverage in the United States: 2006.* U.S. Department of Commerce. Washington D.C.: U.S. Government Printing Office. Available online http://www.census.gov/prod/2007pubs/p60-233.pdf

33. Kaiser Family Foundation. 2007, March. "How changes in medical technology affect health care costs." *Snapshots.* Retrieved September 13, 2007 from http://www.kff.org/insurance/snapshot/chcm030807oth.cfm

34. United Nations. 2001. *World population ageing: 1950-2050.* Available online at: http://www.un.org/esa/population/publications/worldageing19502050/

35. Hobbs, F. B. n.d. *The elderly population.* The U.S. Census Bureau. Retrieved September 14, 2007 at http://www.census.gov/population/www/pop-profile/elderpop.html

36. Callahan, D. and E. Topinkova. 1998. "Is aging a preventable or curable disease?" *Drugs and Aging* 13(2):93-97.

37. Hippocrates of Iphicrates as cited on *GIGA quotes* at http://www.giga-usa.com/quotes/topics/medicine_t002.htm

38. Bennett-Woods, D. 2007b. "Integrating ethical considerations into funding decisions for emerging technologies." *Journal of Nanotechnology Law and Business* 4(1), online.

39. Bennett-Woods, D. 2006. "Nanotechnology in Medicine: Implications of Converging Technologies on the Human Community." *Development* 49(4):54-59.

40. Bennett-Woods, D. and E. Fisher. 2004. *Nanotechnology and the IRB: Toward a new paradigm for analysis and dialogue.* Joint meeting of the European Association for the Study of Science and Technology, and Society for Social Studies of Science. Paris, France. Conference theme: Public Proofs: Science Technology, and Democracy. Paper available online: http://www.csi.ensmp.fr/csi/4S/index.php

41. Clift, R. 2006. "Risk Management and Regulation in an Emerging Technology." In G. Hunt and M. Mehta, eds. *Nanotechnology: Risk, ethics, and law* 140-153. London: Earthscan.

42. Sweet, L. and B. Strohm. 2006. "Nanotechnology: Life-cycle risk management." *Human and Ecological Risk Assessment* 12:528-551.

43. Colvin, V. L. 2003. "The potential environmental impact of engineered nanomaterials." *Nature Biotechnology* 21(10):1166-1170.

44. Williams, R. T. and J. C. Cook. 2007. "Exposure to pharmaceuticals present in the environment." *Drug Information Journal 2007* (online). Drug Information Association. ProQuest Information and Learning Company. Available: http://findarticles.com/p/articles/mi_qa3899/is_200704/ai_n19197488/print

45. President's Commission for the Ethical Study of Problems in Medicine and Biomedical and Behavioral Research. 1983. *Securing access to health care: The ethical implications of differences in the availability of health services.* Volume One. Published report. Washington D.C. Available online at http://www.bioethics.gov/reports/past_commissions/securing_access.pdf

46. Daar, A.S., H. Thorsteinsdóttir, D. K. Martin, A. C. Smith, S. Nast, and P. A. Singer. 2002. "Top ten biotechnologies for improving health in developing countries." *Nature Genetics* 32:229-232.

47. Meridian Institute. 2005. *Nanotechnology and the poor: Opportunities and risks,* published report, Meridian Institute. Available at: http://www.meridian-nano.org/gdnp/paper.php

48. Hauptman, A. and Y. Sharan. 2005, December. *Envisioned developments in nanobiotechnology: Expert survey.* Interdisciplinary Center for Technology Analysis and Forecasting (ICTAF), Tel-Aviv University. Retrieved April 21, 2006 at http://www.ictaf.tau.ac.il/N2L_expert_survey_results.pdf

CHAPTER 10

1. Unknown Author. "Quotations about Perfection." *The Quote Garden.* Quote Garden.com. Available at: http://www.quotegarden.com/perfection.html

2. Bennett-Woods. D. 2006. Adapted from a working paper entitled The Human Person in Transition: Emerging Technologies in Medicine. Paper presented in the session entitled *Brave New Technologies.* Annual meeting of the Society of Social Studies in Science (Vancouver, Canada), *November 2006.*

3. For reference, the reader is recommended to related works by Ray Kurzweil, Ramez Naam, Douglas Mulhall, Gregory Stock, Ted Hughes, Ronald Bailey, Ted Sargent, and Joel Garreau.

4. Naisbitt, J. and P. Aburdene. 1990. *Megatrends 2000.* New York: Avon Books.

5. Ebbesen, M. and T. G. Jensen. 2006. "Nanomedicine: Techniques, potentials, and ethical implications." *Journal of Biomedicine and Biotechnology* 2006:1-11.

6. Roco, M. C. 2003. "Nanotechnology: Convergence with modern biology and medicine." *Current Opinion in Biotechnology* 14:337-346.

7. Bennett-Woods, D. 2007. "Anticipating the impact of nanoscience and nanotechnology in health care." In N. M. Cameron and M. E. Mitchell, eds. *Nanoscale: Issues and perspectives for the nano century.* Hoboken, N.J.: John Wiley & Sons.

8. Hughes, J. 2004. *Citizen cyborg: Why democratic societies must respond to the redesigned human of the future.* Cambridge, MA: Westview Press.

9. Naam, R. 2005. *More than human: Embracing the promise of biological enhancement.* New York: Broadway Books.

10. Bailey, R. 2005. *Liberation biology: The scientific and moral case for the biotech revolution.* Amherst, NY: Prometheus.

11. Fukuyama, F. 2002. *Our posthuman future: Consequences of the biotechnology revolution.* New York: Picador.

12. McKibben, B. 2003. *Enough: Staying human in an engineered age.* New York: Henry Holt and Company.

13. President's Council on Bioethics. 2003. *Beyond therapy: Biotechnology and the pursuit of happiness.* Washington D.C: Dana Press.

14. See for example: Roco, M.C. and W. S. Bainbridge, eds. 2003. *Converging technologies for improving human performance: Nanotechnology, biotechnology, information technology, and cognitive science.* World Technology Evaluation Center. An NSF/DOC report available at http://wtec.org/ConvergingTechnologies/Report/NBIC_report.pdf

15. See for example: European Technology Platform NanoMedicine. November 2006. *Nanomedicine: Nanotechnology for health: Strategic research agenda for nanomedicine.* European Communities. Luxembourg: Office for Official Publications of the European Communities. Available at: http://cordis.europa.eu/nanotechnology/nanomedicine.htm#terms

16. Warren, M.A. 1973/2004. "On the moral and legal status of abortion." Reprinted in R. Munson. 2004. *Intervention and reflection: Basic issues in medical ethics* (7th ed.). Belmont.

17. "The Measure of a Man." *Star Trek Next Generation, Season 2: Episode 9.*

18. "Q Who?" Star *Trek Next Generation, Season 2: Episode 16.* First broadcast in 1989.

19. Roco, M.C. and W. S. Bainbridge. 2003. Overview. In M. C. Roco and W. S. Bainbridge, eds. 2003. *Converging technologies for improving human performance: Nanotechnology, biotechnology, information technology, and cognitive science.* 9-22. World Technology Evaluation Center. An NSF/DOC report available at http://wtec.org/ConvergingTechnologies/Report/NBIC_report.pdf

20. World Health Organization. 1948. *Preamble to the Constitution of the World Health Organization as adopted by the International Health Conference.* Retrieved from http://www.searo.who.int/EN/Section898/Section1441.htm on June 28, 2006.

21. Mykytyn, C. E. 2006. "Anti-aging medicine: Predictions, moral obligations, and biomedical intervention." *Anthropological Quarterly* 79(1):5-31.

22. Khushf, G. "An ethic for enhancing human performance through integrative technologies." In M. C. Roco and W. S. Bainbridge, eds. 2003. *Converging technologies for improving human performance: Nanotechnology, biotechnology, information technology, and cognitive science.* 255-278. World Technology Evaluation Center. An NSF/DOC report available at http://wtec.org/ConvergingTechnologies/Report/NBIC_report.pdf

CHAPTER 11

1. See, for example, the following report of the National Science Board. 2006. *Science and Engineering Indicators 2006.* Two volumes. Arlington, VA: National Science Foundation (volume 1, NSB 06-01; volume 2, NSB 06-01A).

2. Mnyusiwalla, A., A. S. Darr, and P. A. Singer. 2003. "Mind the gap: Science and ethics in nanotechnology." *Nanotechnology* 14:R9-R13.

3. Refer to the NNI website for details on the Societal Dimensions Component Program Area. Available at: http://www.nano.gov/html/society/home_society.html

4. Center for Nanotechnology in Society at Arizona State University, http://cns.asu.edu/index.htm

5. Center for Nanotechnology in Society at University of California at Santa Barbara, http://www.cns.ucsb.edu/

6. University of South Carolina Nanocenter, Nanoscience and Technology Studies, http://nsts.nano.sc.edu/

7. Bond, P. J. April 2, 2004. *A tale of two newspapers: Challenges to nanotechnology development and commercialization.* Speech delivered to the National Nanotechnology Initiative: From Visualization to Commercialization Conference, Washington, D.C. Text available at: http://www.technology.gov/speeches/p_PJB_040402.htm

8. Keiper, A. Summer 2003. "The nanotechnology revolution." *The New Atlantis* (2):17-34.

9. Berube, D. 2006. *Nano-Hype: The truth behind the nanotechnology buzz.* New York: Prometheus.

10. Schummer, J. 2005. "Societal and ethical implications of nanotechnology: Meanings, interest groups and social dynamics." *Techne: Research in Philosophy and Technology* 8(2). Retrieved June 6, 2007 at http://scholar.lib.vt.edu/ejournals/SPT/v8n2/schummer.html

11. Jotterand, F. 2006. "The politization of science and technology." *Journal of Law, Medicine, and Ethics* 34(4):658-666.

12. Fisher, E. 2005. "Lessons learned from the ethical, legal, and social implications program (ELSI): Planning societal implications research for the National Nanotechnology Program." *Technology in Society* 27:321-328.

13. Macoubrie, J. 2005. "Informed public perceptions of nanotechnology and trust in government." Woodrow Wilson International Center for Scholars. Project on Emerging nanotechnologies. Retrieved August 29, 2007 at http://www.wilsoncenter.org/events/docs/macoubriereport.pdf

14. Cobb, M. D., and J. Macoubrie. 2004. "Public perceptions about nanotechnology: Risks, benefits and trust." *Journal of Nanoparticle Research* 6:395-405.

15. Schwartz, P. 1991. *The art of the long view.* New York: Doubleday.

16. See Paul D. Raskin. Global Scenarios: Background Review for the Millennium Ecosystem Assessment, 8 Ecosystems 133, 133-142 (2005); Colin D. Butler et al. Human Health, Well-Being, and Global Ecological Scenarios. 8 Ecosystems 153, 153-162 (2005); Graeme S. Cumming et al. Are Existing Global Scenarios Consistent with Ecological Feedbacks?, 8 Ecosystems 143, 143-152 (2005).

17. Wehrmeyer, W., A. Clayton, and K. Lum. 2002. "Foresighting for Development." *GMI,* 37, Spring: 24-36. Retrieved May 31, 2006 at http://www.greenleaf-publishing.com/content/pdfs/gmi37intr.pdf

18. Berkhout, F. and J. Hertin. 2002. "Foresight Futures Scenarios: Developing and Applying a Participative Strategic Planning Tool." *GMI* 37, Spring: 37-52.

19. Davis, P. K. 2006. "Strategic Planning Amidst Massive Uncertainty in Complex Adaptive Systems: The Case of Defense Planning." In A.A. Minai and Y. Bar-Yam, eds. *Unifying themes in complex systems* 201-214. Berlin: Springer.

20. Bennett-Woods, D. 2007. "Integrating ethical considerations into funding decisions for emerging technologies." *Journal of Nanotechnology Law and Business* 4(1), online.

21. Innes, J. E. and D. E. Booher. 1999. "Consensus building and complex adaptive systems: A framework for evaluating collaborative planning." *Journal of the American Planning Association* 65(14):412-423.

22. Haws, D. R. 2004. "The importance of meta-ethics in engineering education." *Science and Engineering Ethics* 10(2):204-210.

23. Guston, D. H. and D. Sarewitz. 2002. "Real-time technology assessment." *Technology in Society* 24:93-109.

CHAPTER 12

1. Grierson, B. 2007, April 15. "The age of u-turns: Flip-flops get a bad name, but often the course is to reverse course." *Time* 169(16), 74.
2. Senge, P. M. 1990. *The fifth discipline: The art & practice of the learning organization.* New York: Doubleday.
3. Schumacher, E. 1999. Small Is Beautiful. Point Roberts: Hartley & Marks Publishers.
4. Chadwick, R. 2005. "Professional ethics and the 'good' of science." *Interdisciplinary Science Reviews* 30(3):247-256.
5. Mitchell, S. 1988. *Tao Te Ching.* San Francisco: Harper & Row.

Index

Printed and bound by CPI Group (UK) Ltd, Croydon, CR0 4YY

23/10/2024

01777672-0012